J. R. R.

TOLKIEN

AUTHOR OF THE CENTURY

J. R. R.
TOLKIEN

AUTHOR OF THE CENTURY

⊷⟾◉⟾⊷

T. A. Shippey

HOUGHTON MIFFLIN COMPANY
Boston   New York

First Houghton Mifflin paperback edition 2002

For information about permission to reproduce selections
from this book, write to Permissions, Houghton Mifflin Company,
215 Park Avenue South, New York, New York 10003.

Visit our Web site: www.houghtonmifflinbooks.com.

Library of Congress Cataloging-in-Publication Data
Shippey, T. A.
J.R.R. Tolkien : author of the century / T.A. Shippey.
p.   cm.
Originally published: London : HarperCollins, 2000.
Includes bibliographical references and index.
ISBN 0-618-12764-X
ISBN 0-618-25759-4 (pbk.)
1. Tolkien, J. R. R. (John Ronald Reuel), 1892–1973—Criticism
and interpretation. 2. Fantasy fiction, English—History and
criticism. 3. Middle Earth (Imaginary place)
I. Title.
PR6039.O32 Z8238   2001
828'.91209—dc21   2001016973

Printed in the United States of America

QUM 10 9 8 7 6 5 4 3 2 1

LIST OF CONTENTS

117153

ACKNOWLEDGEMENTS

I am grateful to Patrick Curry, Verlyn Flieger, John Garth, Carl Hostetter, Jane Johnson, John Marino, Charles Noad, and Joseph Pearce, who have read sections of this book in draft, sent me advance copies of their own work, preserved me from many errors, and much improved what remains. For any errors of fact or judgement still present I am wholly responsible. I am grateful also to Greenwood Press (an imprint of Greenwood Publishing Group, Inc., Westport, CT), who have allowed me to re-use some of the material in my article in their collection *J.R.R. Tolkien and his Literary Resonances: Views of Middle-earth*, edited by George Clark and Dan Timmons; to Turku University Press, who have given the same permission as regards parts of my article in *Proceedings of the Tolkien Phenomenon*, edited by Keith Battarbee; and to the editors of the volumes just mentioned.

AUTHOR OF THE CENTURY

Fantasy and the fantastic

The dominant literary mode of the twentieth century has been the fantastic. This may appear a surprising claim, which would not have seemed even remotely conceivable at the start of the century and which is bound to encounter fierce resistance even now. However, when the time comes to look back at the century, it seems very likely that future literary historians, detached from the squabbles of our present, will see as its most representative and distinctive works books like J.R.R. Tolkien's *The Lord of the Rings*, and also George Orwell's *Nineteen Eighty-Four* and *Animal Farm*, William Golding's *Lord of the Flies* and *The Inheritors*, Kurt Vonnegut's *Slaughterhouse-Five* and *Cat's Cradle*, Ursula Le Guin's *The Left Hand of Darkness* and *The Dispossessed*, Thomas Pynchon's *The Crying of Lot-49* and *Gravity's Rainbow*. The list could readily be extended, back to the late nineteenth century with H.G. Wells's *The Island of Dr Moreau* and *The War of the Worlds*, and up to writers currently active like Stephen R. Donaldson and George R.R. Martin. It could take in authors as different, not to say opposed, as Kingsley and Martin Amis, Anthony Burgess, Stephen King, Terry Pratchett, Don DeLillo, and Julian Barnes. By the end of the century, even authors deeply committed to the realist novel have often found themselves unable to resist the gravitational pull of the fantastic as a literary mode.

This is not the same, one should note, as fantasy as a literary genre – of the authors listed above, only four besides Tolkien would find their works regularly placed on the 'fantasy' shelves of bookshops, and 'the fantastic' includes many genres besides fantasy: allegory and parable, fairy-tale, horror and science fiction, modern ghost-story and medieval romance. Nevertheless, the point remains. Those authors of the twentieth century who have spoken most powerfully to and for their contemporaries have for some reason found it necessary to use the metaphoric mode of fantasy, to write about worlds and creatures which we know do not exist, whether Tolkien's 'Middle-earth', Orwell's 'Ingsoc', the remote islands of Golding and Wells, or the Martians and Tralfamadorians who burst into peaceful English or American suburbia in Wells and Vonnegut.

A ready explanation for this phenomenon is of course that it represents a kind of literary disease, whose sufferers – the millions of readers of fantasy – should be scorned, pitied, or rehabilitated back to correct and proper taste. Commonly the disease is said to be 'escapism': readers and writers of fantasy are fleeing from reality. The problem with this is that so many of the originators of the later twentieth-century fantastic mode, including all four of those first mentioned above (Tolkien, Orwell, Golding, Vonnegut) are combat veterans, present at or at least deeply involved in the most traumatically significant events of the century, such as the Battle of the Somme (Tolkien), the bombing of Dresden (Vonnegut), the rise and early victory of fascism (Orwell). Nor can anyone say that they turned their backs on these events. Rather, they had to find some way of communicating and commenting on them. It is strange that this had, for some reason, in so many cases to involve fantasy as well as realism, but that is what has happened.

The continuing appeal of Tolkien's fantasy, completely un-expected and completely unpredictable though it was, cannot then be seen as a mere freak of popular taste, to be dismissed or ignored by those sufficiently well-educated to know better. It deserves an explanation and a defence, which this book tries to supply. In the process, I argue that his continuing appeal rests not on mere charm or strangeness (though both are there and can again to some extent be explained), but on a deeply serious response to what will be seen in the end as the major issues of his century: the origin and nature of evil (an eternal issue, but one in Tolkien's lifetime terribly re-focused); human existence in Middle-earth, without the support of divine Revelation; cultural relativity; and the corruptions and continuities of language. These are themes which no one can afford to despise, or need be ashamed of studying. It is true that Tolkien's answers will not appeal to everyone, and are wildly at odds with those given even by many of his contemporaries as listed above. But the first qualification applies to every author who has ever lived, and the second is one of the things that make him distinctive.

However, one of the other things that make him distinctive is his professional authority. On some subjects Tolkien simply knew more, and had thought more deeply, than anyone else in the world. Some have felt (and said) that he should have written his results up in academic treatises instead of fantasy fiction. He might then have been taken more seriously by a limited academic audience. On the other hand, all through his lifetime that academic audience was shrinking, and has now all but vanished. There is an Old English proverb that says (in Old English, and with the usual provocative Old English obscurity), *Ciggendra gehwelc wile þæt hine man gehere*, 'Everyone who cries out wants to be heard!' (Here and in a few places later on, I use the old runic letters þ, ð and ȝ. The first usually represents 'th' as in 'thin', the second 'th' as in 'then'. Where the third is used in this book, it represents -ȝ at the end of a word, -gh- in the middle of one.)

Tolkien wanted to be heard, and he was. But what was it that he had to say?

Tolkien's life and work

For a full account of Tolkien's life, one should turn to Humphrey Carpenter's authorized *Biography* of 1977 (full references to this and other works briefly cited in the text can be found on pp. 329–36 below). But one could sum it up by noting Carpenter's surprising turn on p. 111: 'And after this, you might say, nothing else really happened.' The turning-point Carpenter refers to as 'this' was Tolkien's election to the Rawlinson and Bosworth Chair of Anglo-Saxon in Oxford University in 1925, when he was only thirty-three. The exciting events of Tolkien's life – the stuff most biographers draw on – happened before then. He was born in 1892, in Bloemfontein, South Africa, of English parents. He returned to England very soon, but his father died when he was four, his mother (a convert to Roman Catholicism) when he was twelve. He was brought up in and around Birmingham, and saw himself, despite his foreign birth and German-derived name, as deep-rooted in the counties of the English West Midlands. He met his future wife when he was sixteen and she was nineteen, was eventually forbidden by his guardian to see or correspond with her till he was twenty-one, and wrote proposing marriage to her on his twenty-first birthday. They married while he was at Oxford, but immediately on graduation, in 1915, he took up a commission in the Lancashire Fusiliers. He served as an infantry subaltern on the Somme from July to October 1916, and in that year lost two of his closest friends, killed outright or dead of gangrene. He was then invalided out with trench fever, worked for a short while after the War for the *Oxford English Dictionary*, received first a Readership and then a Chair at Leeds University, and then in 1925 the Anglo-Saxon Chair at Oxford.

And after this 'nothing else really happened'. Tolkien did his job, raised his family, wrote his books, pre-eminently *The Hobbit*, which came out in 1937, and *The Lord of the Rings*, published in three volumes in 1954–5. His main purely academic publications were an edition of the medieval romance *Sir Gawain and the Green Knight*, which he co-edited with E.V. Gordon in 1925, and his British Academy lecture on *Beowulf* in 1936, still accepted as the most significant single essay on the poem out of the (literally) thousands written. He retired from his second Oxford Chair in 1959 (having transferred from the Chair of Anglo-Saxon to the Merton Chair of English Language in 1945). He remained all his life a committed Christian and Catholic, and died, two years after his wife, in 1973. No extra-marital affairs, no sexual oddities, no scandals, strange accusations, or political involvements – nothing, in a way, for a poor biographer to get his teeth into. But what that summary misses out (as Carpenter recognizes) is the inner life, the life of the mind, the world of Tolkien's work, which was also – he refused to distinguish the two – his hobby, his private amusement, his ruling passion.

If Tolkien had ever been asked to describe himself in one word, the word he would have chosen, I believe, would be 'philologist' (see, for instance, the various remarks made in Carpenter's edition of Tolkien's *Letters*, especially p. 264). Tolkien's ruling passion was philology. This is a word which needs some explanation. I have to state here strong personal involvement. I attended the same school as Tolkien, King Edward's, Birmingham, and followed something like the same curriculum. In 1979 I succeeded to the Chair of English Language and Medieval Literature at Leeds which Tolkien had vacated in 1925. I confess that I eventually abolished at Leeds the syllabus which Tolkien had set up two generations before, though I think that in the circumstances of the 1980s I got a deal which Tolkien would himself have reluctantly approved. In between Birmingham and Leeds I had spent seven years as a member of the English faculty at

Oxford, teaching again almost exactly the same curriculum as Tolkien. We were both enmeshed in the same academic duties, and faced the same struggle to keep language and philology on the English Studies curriculum, against the pressing demands to do nothing but literature, post-medieval literature, the relevant, the realistic, the canonical (etc.). There may accordingly be a certain note of factionalism in what I have to say about philology: but at least Tolkien and I were members of the same faction.

In my opinion (it is one not shared, for instance, by the definitions of the *Oxford English Dictionary*), the essence of philology is, first, the study of historical forms of a language or languages, including dialectal or non-standard forms, and also of related languages. Tolkien's central field of study was, naturally, Old and Middle English, roughly speaking the forms of English which date from 700 AD to 1100 (Old) and 1100 to 1500 (Middle) – Old English is often called 'Anglo-Saxon', as in the title of Tolkien's Chair, but Tolkien avoided the term. Closely linked to these languages, however, was Old Norse: there is more Norse in even modern English than people realize, and even more than that in Northern dialects, in which Tolkien took a keen interest. Less closely linked linguistically, but historically connected, are the other ancient languages of Britain, especially Welsh, which Tolkien also studied and admired.

However, philology is not and should not be confined to language study. The texts in which these old forms of the language survive are often literary works of great power and distinctiveness, and (in the philological view) any literary study which ignores them, which refuses to pay the necessary linguistic toll to be able to read them, is accordingly incomplete and impoverished. Conversely, of course, any study which remains solely linguistic (as was often the case with twentieth century philology) is throwing away its best material and its best argument for existence. In philology, *literary and linguistic study are indissoluble*. They ought to be the same thing. Tolkien said exactly that in his letter of application for the Oxford Chair in 1925 (see *Letters*, p. 13), and he pointed to the Leeds curriculum he had set up as proof that he meant it. His aim, he declared, would be:

> to advance, to the best of my ability, the growing neighbour- liness of linguistic and literary studies, which can never be enemies except by misunderstanding or without loss to both; and to continue in a wider and more fertile field the encouragement of philological enthusiasm among the young.

Tolkien was wrong about the 'growing neighbourliness', and about the 'more fertile field', but that was not his fault. If he had been right, he might not have needed to write *The Lord of the Rings*.

Tolkien's fiction is certainly rooted in philology as defined above. He said so himself as forcefully as he could and on every available opportunity, as for instance (*Letters*, p. 219) in a 1955 letter to his American publishers, trying to correct impressions given by a previous letter excerpted in the *New York Times*:

> the remark about 'philology' [in the excerpted letter, 'I am a philologist, and all my work is philological'] was intended to allude to what is I think a primary 'fact' about my work, that it is all of a piece, and *fundamentally linguistic* in inspi- ration ... The invention of languages is the foundation. The 'stories' were made rather to provide a world for the languages than the reverse. To me a name comes first and the story follows.

The emphasis in the passage quoted is Tolkien's, and he could hardly have put what he said more strongly, but his declaration has been met for the most part by bafflement or denial. And there is a respectable reason for this (along with many less respectable ones), for Tolkien was the holder of several highly personal if not heretical views about language. He thought that people, and perhaps as a result of their confused linguistic history especially English people, could detect historical strata in language without knowing how they did it. They knew that names like Ugthorpe and Stainby were Northern without knowing they were Norse; they knew Winchcombe and Cumrew must be in the West without recognizing that the word *cŵm* is Welsh. They could feel linguistic style in words. Along with this, Tolkien believed that languages could be intrinsically attractive, or intrinsically repulsive. The Black Speech of Sauron and the orcs is repulsive. When Gandalf uses it in 'The Council of Elrond', 'All trembled, and the Elves stopped their ears'; Elrond rebukes Gandalf for using the language, not for what he says in it. By contrast Tolkien thought that Welsh, and Finnish, were intrinsically beautiful; he modelled his invented Elf-languages on their phonetic and grammatical patterns, Sindarin and Quenya respectively. It is a sign of these convictions that again and again in *The Lord of the Rings* he has characters speak in these languages *without bothering to translate them*. The point, or a point, is made by the sound alone – just as allusions to the old legends of previous ages say something without the legends necessarily being told.

But Tolkien also thought – and this takes us back to the roots of his invention – that philology could take you back even beyond the ancient texts it studied. He believed that it was possible sometimes to feel one's way back from words as they survived in later periods to concepts which had long since vanished, but which had surely existed, or else the word would not exist. This process was made much more plausible if it was done comparatively (philology only became a science when it became *comparative* philology). The word 'dwarf' exists in modern English, for instance, but it was originally the same word as modern German *Zwerg*, and philology can explain exactly how they came to differ, and how they relate to Old Norse *dvergr*. But if the three different

languages have the same word, and if in all of them some fragments survive of belief in a similar race of creatures, is it not
legitimate first to 'reconstruct' the word from which all the later
ones must derive – it would have been something like *dvairgs*
– and then the concept that had fitted it? [The asterisk before
dvairgs is the conventional way of indicating that a word has
never been recorded, but must (surely) have existed, and there
is of course enormous room for error in creating *-words, and
*-things.] Still, that is the way Tolkien's mind worked, and many
more detailed examples are given later on in this book. But
the main point is this. However fanciful Tolkien's creation of
Middle-earth was, he did not think that he was *entirely* making
it up. He was 'reconstructing', he was harmonizing contradictions
in his source-texts, sometimes he was supplying entirely new
concepts (like hobbits), but he was also reaching back to an
imaginative world which he believed had once really existed, at
least in a collective imagination: and for this he had a very great
deal of admittedly scattered evidence.

Tolkien furthermore had distinguished predecessors in the previous century. In the 1830s Elias Lönnrot, the Finn, put together
what is now the Finnish national epic, the *Kalevala*, from scattered
songs and lays performed for him by many traditional singers;
he 'reconstructed', in fact, the connected poem which he believed
(probably wrongly) had once existed. At much the same time Jacob
and Wilhelm Grimm, in Germany, took up their enormous project
of compiling at once a German grammar, a German dictionary,
a German mythology, a German cycle of heroic legends, and
of course a corpus of German fairy-tales – literary and linguistic
study pursued without distinction, just as it should be. In Denmark Nikolai Grundtvig had set himself to re-creating Danish
national identity, with passionate attention to the saga and epic
literature of old, as to the ballad-literature of later periods, eventually brought together by his son Sven. But in England there had
been no such nineteenth-century project. When Tolkien then

said, as he did (see *Letters*, p. 144), that he had once hoped 'to make a body of more or less connected legend' which he could dedicate simply 'to England; to my country', he was not saying something completely unprecedented; though he did admit ruefully, in 1951, that his hopes had shrunk. Ten years later he might have felt much closer to success.

Tolkien, then, was a philologist before he was a mythologist, and a mythologist, at least in intention, before he ever became a writer of fantasy fiction. His beliefs about language and about mythology were sometimes original and sometimes extreme, but never irrational, and he was able to express them perfectly clearly. In the end he decided to express them not through abstract argument, but by demonstration, and the success of the demonstration has gone a long way to showing that he did often have a point: especially in his belief, which I share, that a taste for philology, for the history of language in all its forms, names and place-names included, is much more widespread in the population at large than educators and arbiters of taste like to think. In his 1959 'Valedictory Address to the University of Oxford' (reprinted in *Essays*, pp. 224–40), Tolkien concluded that the problem lay not with the philologists nor with those they taught but with what he called 'misologists' – haters of the word. There would be no harm in them if they simply concluded language study was not for them, out of dullness or ignorance. But what he felt, Tolkien said, was:

> grievance that certain professional persons should suppose their dullness and ignorance to be a human norm; and anger when they have sought to impose the limitation of their minds upon younger minds, dissuading those with philological curiosity from their bent, encouraging those without this interest to believe that their lack marked them as minds of a superior order.

Behind this grievance and this anger was, of course, failure and defeat. It is now very hard to pursue a course of philology of the kind Tolkien would have approved in any British or American university. The misologists won, in the academic world; as did the realists, the modernists, the post-modernists, the despisers of fantasy.

But they lost outside the academic world. It is not long since I heard the commissioning editor of a major publishing house say, 'Only fantasy is mass-market. Everything else is cult-fiction.' (Reflective pause.) 'That includes mainstream.' He was defending his own buying strategy, and doubtless exaggerating, but there is a good deal of hard evidence to support him. Tolkien cried out to be heard, and we have still to find out what he was saying. There should be no doubt, though, that he found listeners, and that they found whatever he was saying worth their while.

The *author of the century*

After this preamble, one may now consider the claim, or claims, made in this book's title. Can Tolkien be said to be '*the* author of the century'? Any such claim, ambitious as it is, could rest on three different bases. The first of them is simply democratic. That is what opinion polls, and sales figures, appear to show. The details are given immediately below, along with some consideration of how they should be interpreted, and how they have been; but one can say without qualification that a large number of readers, both in Britain and internationally, have agreed with the claim, and that they have done this furthermore without prompting or direction.

The second argument is generic. As the commissioning editor said, fantasy, especially heroic fantasy, is now a major commercial genre. It existed before Tolkien, as is again discussed below, and it is possible to say that it would have existed, and would have developed into the genre it has become, without the lead of *The Lord of the Rings*. This seems, however, rather doubtful. When it came out in 1954–5 *The Lord of the Rings* was quite clearly a sport, a mutation, *lusus naturae*, a one-item category on its own. One can only marvel, looking back, on the boldness and determination of Sir Stanley Unwin in publishing it at all – though significantly enough, he hedged his bet by entering into a profit-sharing agreement with Tolkien by which Tolkien got nothing till there were some profits to share, a matter clearly of some doubt at the time. Unwin had moreover continued to support and encourage his author over a seventeen-year gestation period which in the event delivered quite a different birth from what had been intended. It is true that he never had to pay over the large sums which James Joyce's backers did, for instance, while Joyce was producing *Ulysses*, but then neither he nor Tolkien ever had the kind of support from a professional literary élite which Joyce and his benefactors could count on. However, while *Ulysses* has had few direct imitators, though many admirers, after *The Lord of the Rings* the heroic fantasy 'trilogy' became almost a standard literary form. Any bookshop in the English-speaking world will now have a section devoted to fantasy, and very few of the works in the section will be entirely without the mark of Tolkien – sometimes branded deep in style and layout, sometimes showing itself in unconscious assumptions about the nature and personnel of the authors' invented fantasy worlds. The imitations, or emulations, naturally vary very widely in quality, but they all give pleasure to someone. One of the things that Tolkien did was to open up a new continent of imaginative space for many millions of readers, and hundreds of writers – though he himself would have said (see above) that it was an old continent which he was merely rediscovering. An acceptably philological way of putting it might be to say that Tolkien was the Chrétien de Troyes of the twentieth century. Chrétien, in the twelfth century, did not *invent* the Arthurian romance, which must have existed in some form before his time, but he showed what could be done with it; it is a genre whose potential has never been exhausted in the eight centuries since. In the same way, Tolkien did not *invent* heroic fantasy, but he showed what could be done with it; he established a genre whose durability we cannot estimate.

The third argument has to be qualitative. Popularity does not guarantee literary quality, as everybody knows, but it never comes about for no reason. Nor are those reasons always and necessarily feeble or meretricious ones, though there has long been a tendency among the literary and educational élite to think so. To give just one example, in my youth Charles Dickens was not regarded as a suitable author for those reading English Studies at university, because for all his commercial popularity (or perhaps because of his commercial popularity) he had been downgraded from being 'a novelist' to being 'an entertainer'. The opinion was reversed as critics developed broader interests and better tools; but although critical interest has stretched to include Dickens, it has not for the most part stretched to include Tolkien, and is still uneasy about the whole area of fantasy and the fantastic – though this includes, as has been said, many of the most serious and influential works of the whole of the later twentieth century, and its most characteristic, novel and distinctive genres (such as science fiction).

The qualitative case for these genres, including the fantasy genre, needs to be made, and the qualitative case for Tolkien must be a major part of it. It is not a particularly difficult case to make, but it does require a certain open-mindedness as to what people are allowed to get from their reading. Too many critics have defined 'quality' in such a way as to exclude anything other than what they have been taught to like. To use the modern jargon, they 'privilege' their own assumptions and prejudices, often class-prejudices, against the reading choices of their fellow-men and fellow-women, often without thinking twice about it. But many people have been deeply and lastingly moved by

Tolkien's works, and even if one does not share the feeling, one should be able to understand why.

In the following sections, I consider further the first two arguments outlined above, and set out the plan and scope of the chapters which follow, which form in their entirety my expansion of the third argument, about literary quality; and my answer to the question about what Tolkien felt he had to say.

Tolkien and the polls

Tolkien's sales figures have always been an annoyance for his detractors, and as early as the 1960s commentators had been predicting that they would soon fall, or declaring that they had started to fall, so that the whole 'cult' or 'craze' would pass or was already passing into 'merciful oblivion' (so Philip Toynbee wrote in the *Observer* on 6th August 1961), just like flared jeans or hula hoops. The commentators were wrong about this – a surprise in itself, since Tolkien never followed up with either a *Hobbit*-sequel for the children's market nor a *Lord of the Rings*-sequel for the adult market. But the whole issue of his continuing popularity was brought forward dramatically during 1997.

Very briefly – there is a more extensive account in Joseph Pearce's book of 1998, *Tolkien: Man and Myth*, to which I am indebted – late in 1996 Waterstone's, the British bookshop chain, and BBC Channel Four's programme *Book Choice* decided between them to commission a readers' poll to determine 'the five books you consider the greatest of the century'. Some 26,000 readers replied, of whom rather more than 5,000 cast their first place vote for J.R.R. Tolkien's *The Lord of the Rings*. Gordon Kerr, the marketing manager for Waterstone's, said that *The Lord of the Rings* came consistently top in almost every branch in Britain (105 of them), and in every region except Wales, where James Joyce's *Ulysses* took first place. The result was greeted with horror among professional critics and journalists, and the *Daily Telegraph* decided accordingly to repeat the exercise among its readers, a rather different group. Their poll produced the same result. The Folio Society then confirmed that during 1996 it had canvassed its entire membership to find out which ten books the members would most like to see in Folio Society editions, and had got 10,000 votes for *The Lord of the Rings*, which came first once again. 50,000 readers are said to have taken part in a July 1997 poll for the television programme *Bookworm*, but the result was yet again the same. In 1999 the *Daily Telegraph* reported that a Mori poll commissioned by the chocolate firm Nestlé had actually managed to get a different result, in which *The Lord of the Rings* (at last) only came second! But the top spot went to the Bible, a special case, and also ineligible for the twentieth-century competition which had begun the sequence.

These results were routinely and repeatedly derided by professional critics and journalists (the latter group, of course, often the products of university literature departments). Joseph Pearce opens his book with Susan Jeffreys, of the *Sunday Times*, who on 26th January 1997 reported a colleague's reaction to the news that *The Lord of the Rings* had won the BBC/Waterstone's poll as: 'Oh hell! Has it? Oh my God. Dear oh dear. Dear oh dear oh dear'. This at least sounds sincere, if not deeply thoughtful; but Jeffreys reported also that the reaction 'was echoed up and down the country wherever one or two literati gathered together'. She meant, surely, 'two or three literati', unless the literati talk only to themselves (a thought that does occur); and the term *literati* is itself interesting. It clearly does not mean 'the lettered, the literate', because obviously that group includes the devotees of *The Lord of the Rings*, the group being complained about (they couldn't be devotees if they couldn't read). In Jeffreys's usage, *literati* must mean 'those who know about literature'. And those who know, of course, know what they are supposed to know. The opinion is entirely self-enclosed.

Other commentators meanwhile suggested that the first poll by Waterstone's must have been influenced by concerted action on the part of the Tolkien Society. The Society denies this, and points out that even if every one of their five hundred members *had* voted, this would still have been less than the margin of victory (1,200 votes) over the runner-up, George Orwell's *Nineteen Eighty-Four*. Germaine Greer took another tack by declaring angrily in the Winter/Spring 1997 issue of *W: the Waterstone's Magazine*, that ever since her arrival at Cambridge in 1964, 'it has been my nightmare that Tolkien would turn out to be the most influential writer of the twentieth century. The bad dream has materialized'. She added, 'The books that come in Tolkien's train are more or less what you would expect; flight from reality is their dominating characteristic'. It seems strange to see novels like *Nineteen Eighty-Four* and fables like *Animal Farm* castigated for 'flight from reality', though of course they are not novels of mainstream realism: as I remark above, it seems that some themes, including public and political ones, are best handled as fable or as fantasy. And calling something that has after all happened a 'bad dream' does not suggest too strong a grip on reality by the critic. Tolkien in any case had his own view on the modern development of words like 'reality, real, realist, realistic', see p. 76 below: Saruman, the collaborator, the wizard who goes over to the other side because it seems the stronger, would no doubt have called himself a 'realist', though that would not make him one.

It remains perfectly sensible, of course, to say that popular polls are no guide to literary value, any more than sales figures, and indeed both statements are no doubt true. The figures ought however to have produced some sort of considered response, even explanation, from professional critics of literature, rather than the nettled outrage which they got. To quote the critic Darko Suvin (writing primarily about science fiction, but extending his point to all forms of 'paraliterature' or commercial literary production):

a discipline which refuses to take into account 90 per cent or more of what constitutes its domain seems to me not only to have large zones of blindness but also to run serious risks of distorted vision in the small zone it focuses on (so-called high lit.)

(Suvin, 1979, p. vii)

This 'noncanonic, repressed twin of Literature', he adds, is 'the literature that is really read – as opposed to most literature taught in schools'. And this indicates a further oddity about the polling results above. If one looks at the Waterstone's list overall, it is very easy to detect what a correspondent in the *Times Educational Supplement* called 'the formative influence of school set texts on a nation's reading habits'. Even leaving aside the Welsh preference for Joyce's *Ulysses* – the work most intensely promoted by academics and educationalists – the leading places after *The Lord of the Rings* were taken by Orwell's *Nineteen Eighty-Four* and *Animal Farm*, and Salinger's *Catcher in the Rye*, with Golding's *Lord of the Flies* not far away: all very familiar school set texts, routinely taught and examined, and for the most part comparatively short. *The Lord of the Rings* is however rarely if ever set as a text in schools or universities. Apart from the dislike of the educational establishment, it is too long, at over half a million words. The following it has acquired has all been the result of personal choice, not institutional direction.

A further thought which ought to have struck commentators is this. It is quite possible, as said above, to separate the evidence of mass sales from claims for lasting or literary value. There are several authors now who out-sell Tolkien on an annual basis, or who have done so in the recent past – Barbara Taylor Bradford, Tom Clancy, Catherine Cookson, Michael Crichton, John Grisham, Stephen King – to offer a mere selection from the first half of the alphabet. None of them could achieve their popularity without virtues of some kind, and as Suvin implies above, critical reluctance even to look for these virtues says more about the critics than the popular authors. Just the same, the works of those listed above are not much like Tolkien's. It is in fact hard to think of a work (except perhaps in their different ways *The Silmarillion* and *Finnegans Wake*) written with less concern for commercial considerations than *The Lord of the Rings*. No market researcher in the 1950s could possibly have predicted its success. It was long, difficult, trailed with appendices, studded with quotations in unknown languages which the author did not always translate, and utterly strange. It had, indeed, to create its own market. And two further striking points about it are, first, that it did, and second, that unlike most of the works of the authors mentioned above (to whom I mean no disrespect) it has had a continued shelf-life. *The Hobbit* has stayed in print for more than sixty years, selling over forty million copies, the *Lord of the Rings* for nearly fifty years, selling over fifty million (which, since it is published usually in three-volume format, comes to close on a hundred and fifty million separate sales).

Tolkien and the fantasy genre

To take up my second argument, and to return to the point about creating a market, it would not be true to say that there was no such thing as epic fantasy before Tolkien: there was a tradition of English and Irish writers before him, such as E.R. Eddison and Lord Dunsany, and a parallel tradition also of American writers appearing in pulp-magazines such as *Weird Tales* and *Unknown*. (I discuss and exemplify these in my anthology *The Oxford Book of Fantasy Stories*, 1994). *The Lord of the Rings* however altered reading tastes rapidly and lastingly. Several hundred English-language fantasy novels are currently being published annually. The influence of Tolkien on them is often apparent from their titles – I note the 'Malloreon' sequence by David

Eddings, whose first title is *The Guardians of the West*, with *The Fellowship of the Talisman*, *The Halfling's Gem* and *Lúthien's Quest* coming from other authors. Most writers do better at concealing their literary ancestry, but the first works even of authors who have found their own highly distinctive voices, like Stephen Donaldson or Alan Garner, habitually betray deep Tolkienian influence, as is discussed at greater length below (see pp. 321–4). Terry Pratchett, whose works have now been reliable best-sellers for almost twenty years, began with what is obviously in part an affectionate parody of Tolkien (and of other fantasy writers), *The Colour of Magic*. Tolkien furthermore provided much of the inspiration, the personnel and the material, for early fantasy games and for role-playing games of the 'Dungeons and Dragons' type: the article on 'Fantasy Games' in John Clute and John Grant's *Encyclopedia of Fantasy* lists, among others, *Battle of Helm's Deep*, *Siege of Minas Tirith*, and *The Middle Earth Role Playing System*. Spin-offs from these into computer games are still developing and multiplying. Middle-earth became a cultural phenomenon, a part of many people's mental furniture.

Nor were these admirers, despite what Tolkien's critics have said, simply uneducated or retarded. The division in tastes was never between low/popular and high/educated, it lay rather between generally-educated and professionally-educated. It appears that people have to be educated *out* of a taste for Tolkien rather than into it. Some, of course, say that that is what education is supposed to do, 'lead out rather than put in', to quote the familiar educationalists' motto. Tolkien would have replied that he was satisfying a taste – the taste for fairy-tale – which is natural to us, which goes back as far as we have written records of any sort, to the Old Testament and Homer's *Odyssey*, and which is found in all human societies. If our arbiters of taste insist that this taste should be suppressed, then it is they who are flying from reality. As proper *literati* might put it, *Naturam expelles furca, tamen usque recurret* – Latin for, 'you can chuck out nature with a pitchfork, but it'll come back just the same'.

An *author of the twentieth century*

The creation, or re-creation, of a whole publishing genre is a strange result for a book written without the slightest commercial awareness; in a style which is frequently professorial; and which appeared as a first adult novel when its author was already sixty-two (an event not entirely dissimilar, one might note, to the appearance of Joyce's *Ulysses* as a first and last major work when its author was forty).

Whatever one thinks of the last parallel (and there are other parallels between Joyce and Tolkien which might be drawn, see pp. 310–14 below), there can at least be no doubt that – to sum up what has been said above – *The Lord of the Rings* has established itself as a lasting classic, without the help and against the active hostility of the professionals of taste; and has furthermore largely created the expectations and established the conventions of a new and flourishing genre. It and its author deserve more than the routine and reflexive dismissals (or denials) which they have received. *The Lord of the Rings*, and *The Hobbit*, have said something important, and meant something important, to a high proportion of their many millions of readers. All but the professionally incurious might well ask, what? Is it something timeless? Is it something contemporary? Is it (and it is) both at once?

This book attempts accordingly to explain Tolkien's success and to make out the case for his importance. It follows my earlier book on Tolkien, *The Road to Middle-earth* (1982, revised edition 1992), but with several differences of emphasis and of understanding. The main one is that *The Road to Middle-earth* was to some considerable extent a work of professional piety – using piety in the old sense of respect for one's forebears or predecessors. In it my concern was above all to set Tolkien's work in a philological context, as outlined above, but in much greater detail. I still feel that the piety was justified, and that the point needed to be made. However, in the first place I have reluctantly to concede that not everyone takes to Gothic, or even (in extreme cases) to Old Norse. Moreover, even professional linguists accept that while one can study language 'diachronically', i.e. historically, across time, there is also much to be gained by studying it 'synchronically', i.e. as it exists at any given moment. In the same way, while I remain convinced that Tolkien cannot be properly discussed without some considerable awareness of the ancient works and the ancient world which he tried to revive (awareness which I try to promote in the following chapters), I now accept that he needs also to be looked at and interpreted within his own time, as *an* 'author of the century', the twentieth century, responding to the issues and the anxieties of that century. This latter is the way that most people read him, and it is only reasonable to try to follow suit.

Plan and scope of this book

The six main chapters which follow try accordingly not only to discuss Tolkien's many sources of inspiration for 'Middle-earth', but also to show why Middle-earth has been a vital contemporary inspiration for so many readers. They are in one sense *not* chronological. We now know – as we did not when I wrote the first version of *The Road to Middle-earth* – that Tolkien spent most of his life working on the complex of legends which eventually appeared, posthumously, as *The Silmarillion*, the *Unfinished Tales*, and the twelve-volume *History of Middle-earth*. Much of this existed before the writing of *The Hobbit* and *The Lord of the Rings*, he turned back to it during the long composition of both works, and again after they were published. If one were tracing

Tolkien's own development as an author, it would make sense to start from the beginning, and to treat *The Hobbit* and *The Lord of the Rings* as the offshoots which in a way they are. However, if one is considering his impact on and his relationship to his own time, the influential works are clearly the two of the hobbit-sequence, and I accordingly begin with them.

In chapter I I consider in particular the literary function of hobbits, and of Bilbo Baggins, their representative. I argue that they are above all anachronisms, creatures of the early modern world of Tolkien's youth drawn, like Bilbo, into the far more archaic and heroic world of dwarves and dragons, wargs and were-bears. However Tolkien, as a philologist, and also as an infantry veteran, was deeply conscious of the strong continuity between that heroic world and the modern one. Much of the vocabulary of Old English is exactly the same as that of modern English; many of its situations seem to recur. Meanwhile Robert Graves, an almost exact contemporary of Tolkien's, remarks in his 1929 memoir *Goodbye to All That* that when he arrived at Oxford in 1919 his Anglo-Saxon lecturer (one wonders who it was) disparaged his own subject, and said it had no interest or relevance. Graves disagreed. He thought that:

> Beowulf lying wrapped in a blanket among his platoon of drunken thanes in the Gothland billet; Judith going for a *promenade* to Holofernes's staff-tent; and *Brunanburgh* with its bayonet-and-cosh fighting – all this came far closer to most of us than the drawing-room and deer-park atmosphere of the eighteenth century.

Graves's language is deliberately anachronistic: 'platoon', 'billet', 'staff-tent', 'cosh', are all modern words with immediate World War I meanings, while *promenade* is a soldiers' euphemism. 'Thanes' on the other hand is completely archaic. Yet Graves's point is precisely to *deny* any sense of anachronism. In its way –

a much more complex and extensive way – *The Hobbit* carries out the same exercise. It takes its readers, even child readers, into a totally unfamiliar world, but then indicates to them that it is not totally unfamiliar, that they have a birth-right in it of their own. The book operates frequently through a clash of styles – linguistic, moral, behavioural – but ends by demonstrating unity and understanding on a level deeper than style.

With Middle-earth in imaginative existence, it might have been thought relatively easy to produce the sequel which Tolkien's publisher immediately requested. Chapter II deals with Tolkien's problems in creating *The Lord of the Rings*, both of invention and of organization, problems which have become much clearer with the publication of much of his early drafts. The drafts are almost dismaying to enthusiasts, for one of the things they reveal is that the neat thematic patterns recognized by so many critics (myself included) seem always to have been afterthoughts. When he started writing Tolkien had literally no idea at all of where he was going. Yet by the end not only are there unmistakably tight patterns of cultural contrast and cultural parallel, not only is the work marked by continuing deliberate dramatic irony, its entire structure depends on a chronology which Tolkien developed with great care, and printed in his Appendix B. I argue that this is one of the major differences between *The Lord of the Rings* and (as far as I can tell) all its emulators. No professional or commercially-oriented author would ever have tried anything as difficult or as demanding of its readers' attention. Yet Tolkien, both in overall organization and in the organization of major sections like the chapter 'The Council of Elrond', successfully presented an immensely complex pattern of narrative 'interlace' – which works, like the best narrative strategies, even on those unconscious of it, but which nevertheless deserves proper appreciation.

Chapters III and IV take up the two most immediately contemporary themes in *The Lord of the Rings* – evil, and myth. As was again remarked above, it is possible to see Tolkien as one of a group of 'traumatized authors', all of them extremely influential (they mostly rank high in polls like Waterstone's), all of them tending to write fantasy or fable. The group includes, besides the names mentioned on p. viii (Tolkien, Orwell, Golding, Vonnegut), others such as Tolkien's friend C.S. Lewis, T.H. White, and Joseph Heller. Their experiences include being shot (Orwell and Lewis were both all but fatally wounded on the battlefield), and being bombed (Vonnegut was actually in Dresden the night it was destroyed). Ursula Le Guin, though without similarly direct experience of violence, is the daughter of Theodora Kroeber, who wrote three different accounts of 'Ishi', the last survivor of the eventually total elimination of the Yahi Indians of California. Most of these authors, then, had close or even direct first-hand experience of some of the worst horrors of the twentieth century, horrors which did not and could not exist before it: the Somme, Guernica, Belsen, Dresden, industrialized warfare, genocide.

Their very different but related experiences left all of them, one may say, with an underlying problem. They were bone-deep convinced that they had come into contact with something irrevocably evil. They also – like Graves in the quotation above, but far more seriously – felt that the explanations for this which they were given by the official organs of their culture were hopelessly inadequate, out of date, at best irrelevant, at worst part of the evil itself. Orwell returned from Spain to find his own personal experience, including being shot, dismissed as a non-event, a political aberration. Vonnegut spent twenty years wondering how he could write about the central event of his life, the destruction of Dresden, in a way that could possibly be appreciated, while dealing with people who preferred to deny or ignore it. By contrast the dominant moral philosophers of these authors' time and culture included people like Bertrand Russell (an author, like Tolkien, published by Stanley Unwin, and according to his 1967 festschrift, the 'philosopher of the century'). But what could Russell tell Lewis, say, about what he had experienced in Flanders?

In World War I Russell was a pacifist: an honourable stance, but for 'traumatized authors' not a helpful one, and as Russell came painfully to realize at the outbreak of World War II, in some circumstances an untenable one. One of the aspects of the trauma for the authors I have mentioned was that when it came to finding explanations, they were on their own.

All of them responded with highly individual images, and theories, of evil. I mention here only Le Guin's 'The Ones who Walk Away from Omelas' (a civilization which rests on the torture of an idiot child); Orwell's interrogator-figure O'Brien (the future as a boot stamping on a human face, for ever); White's *Book of Merlyn* (humanity redefined not as *homo sapiens* but as *homo ferox*). Obviously the list could be extended. In Tolkien's case I see his central image of evil as that of the 'wraith', an old word, but one which has been given terrible new force. Round this ambiguous image there revolves the concept of the Ring, which itself embodies two distinct and competing theses about the nature of evil, the one officially accepted (but hard to credit), the other threateningly heretical (but all too easy, in modern circumstances, to accept). Tolkien not only poses questions about evil, he also provides answers and solutions – one of the things which has made him unpopular with the professionally gloomy or fashionably nihilist. Nevertheless, although his concern and the concern of the authors I mention is not with the private and the personal (the themes of the 'modernist' novel), but with the public and the political, it should be obvious that to all but the sheltered classes of this century, the most important events in private lives (and even more, in deaths) have often *been* public and political. It is those who turn away from that thought, who prefer to remain in what Graves called the 'drawing-room' areas of literary tradition, who are in 'flight from reality'.

Chapter IV extends the discussion of evil to consideration, first, of the evident connections between *The Lord of the Rings* and modern history (Tolkien denied 'allegory' but conceded 'applicability'); and second, of the attempt to reach out beyond contemporary relevance and beyond archaism to something which governs both – timelessness, 'the mythic dimension', and Tolkien's own idiosyncratic but well-informed view of literary tradition. This chapter also takes up one of the major apparent paradoxes of *The Lord of the Rings*. It was written, we know, by a devout and believing Christian, and has been seen by many as a deeply religious work. Yet it contains almost no direct religious reference at all. Returning to the theme taken up in chapter I, I argue that *The Lord of the Rings* can be taken in itself as a myth, in the sense of a work of mediation, reconciling what appear to be incompatibles: heathen and Christian, escapism and reality, immediate victory and lasting defeat, lasting defeat and ultimate victory.

The last two chapters set Tolkien's two major works in the context of his other continuing literary activities, both those published and those unpublished in his own lifetime. A major aim of chapter V is to provide a guide to reading the published *Silmarillion*, a work which is quite outside any modern reading or writing conventions, but which has never received the credit normally extended to the 'experimental'. However, it considers also the growth and development of the (non-italicized) 'Silmarillion', by which I mean the many parts of the overall legendarium eventually published in the twelve-volume sequence, *The History of Middle-earth*. Two dominant ideas in this chapter are, first, Tolkien's own complex notion of literary 'depth', by which a work – like Lord Macaulay's famous *Lays of Ancient Rome* – gains added charm from having a sense behind it of an older history now lost, as well as of a later and less truthful history now more familiar; and second, the deep sadness which infuses all 'Silmarillion' versions, and which may be seen with hindsight to underlie even the cheerful hobbits and their epic, *The Lord of the Rings*.

Chapter VI takes up some of the reasons for this sadness, and considers what some of Tolkien's minor works tell us (and for all his dislike of biography, were intended to tell us) about his inner life. A feature of this chapter is the claim that at least two of his minor published works, 'Leaf by Niggle' and *Smith of Wootton Major*, are in their different ways 'autobiographical allegories'. The case may seem a hard one to make, since Tolkien's expressed disapproval of allegories is well-known. I hope nevertheless to have made it, even within Tolkien's own deliberately narrow definition of allegory. My view is that he felt allegory had its place, and its rules, and that his scorn was reserved for those who insisted on using it and detecting it outside that place. In between my readings of those two works, one early, one late, I consider the small corpus of poems which Tolkien published, and sometimes republished, in his own lifetime, relating them in some cases to his personal myth of 'the Lost Road', expressed in two separate abortive attempts to write another major fiction. Besides *The Hobbit* and *The Lord of the Rings*, Tolkien's only other entirely successful narrative published in his own lifetime was the unusually light-hearted novella *Farmer Giles of Ham*. I attempt further to fit this, along with two other poetic narratives, into Tolkien's again idiosyncratic but well-informed view of literary history.

In the Afterword, finally, I take up once more the criticisms of Tolkien which underlie the outrage mentioned at the start of this Foreword. It is to a large extent a guessing-game. Very few of Tolkien's critics (there are some honourable exceptions) have been prepared to put their dislike into an organized shape which can be debated; one of the most vehement indeed confessed to me, in private, in the lift taking us out of BBC House after a radio debate, that he had never actually read *The Lord of the Rings* which he had just been attacking. I find myself accordingly sometimes making the case against so that I can make the case for, not an ideal procedure. Still, the repeatedly expressed dislike of an influential and easily-identifiable section of the literary world is part of the phenomenon. Very probably the reason for the dislike has a good deal to do with the reasons for the success. Tolkien has challenged the very authority of the *literati*, and this is never forgiven.

The obverse of this exercise is to look in slightly more detail at Tolkien's emulators. We may not be sure exactly what people have liked in Tolkien's work, but we can see what writers have tried to imitate, as also what they have shied away from. Some of them, of course, may have superseded him, used his work only as a starting-point for quite different directions, even in some respects outdone him. It could be said that this latter is one of the best things that can happen to an innovative writer: Tolkien indeed wrote (see *Letters*, p. 145) that he had hoped once that his story-cycles would 'yet leave scope for other minds and hands'. He then immediately and self-deprecatingly dismissed his hope as 'Absurd' (this was in 1951, *The Lord of the Rings* still unpublished).

However, similar results have been achieved by other philologist-creators. Lönnrot's *Kalevala* is now viewed with suspicion by scholars, because Lönnrot, like Walter Scott with the Border Ballads, did not just collect and transcribe, but wrote, rewrote and interpolated, so that you cannot tell what is by him and what is 'authentic'. Just the same, the date of publication of the *Kalevala* remains a national holiday in Finland, and the work has become a cornerstone of national culture. Very similar accusations of interference and meddling have been made about the Grimms and their *Fairy-Tales*; but for two centuries the tales have enriched not just national but international culture, and delighted hundreds of millions of child and adult readers. Nikolai Grundtvig, the Dane, insisted on the concept of *levende ord*, 'the living word'. It is not enough for the philologist, the 'word-lover', to be scholarly. The scholar also has to transmit his results into the life and speech and imagination of the greater world.

In 1951 Tolkien, like Théoden King when we first meet him,

can have had little hope of such success. By his death-day, however, he could well have said that, like Théoden, when he went
to join his (philological) fathers, 'even in their mighty company
I shall not now be ashamed'. Tolkien left a legacy as rich as any
of his predecessors'.

CHAPTER I

⊰⭑⟳⟲⭑⊱

THE HOBBIT:
RE-INVENTING MIDDLE-EARTH

A moment of inspiration?

The story of how J.R.R. Tolkien came to be launched on his career, not as a writer of fiction – this had begun many years before – but as a writer of published fiction, is a familiar one. According to Tolkien's own account, he was sitting one day, after he had become Professor of Anglo-Saxon in the University of Oxford, in his home in Northmoor Road, laboriously marking School Certificate papers: something, one should note, which was no part of his university duties, but which many academics then undertook as a summer-time extra to supplement their incomes. A boring job, then, engaging Tolkien's intellect at well below its top level, but at the same time one which in decency to the candidates had to be done conscientiously, with full alertness: academic piece-work, but piece-work which, unlike sewing or standing on a production line, gave no opportunity for the mind to wander. In this circumstance (the strain of which only those who have marked, say, five hundred handwritten scripts on the same subject will fully appreciate) Tolkien turned over a page to find that a candidate:

had mercifully left one of the pages with no writing on it (which is the best thing that can possibly happen to an examiner) and I wrote on it: *'In a hole in the ground there lived a hobbit.'* Names always generate a story in my mind. Eventually I thought I'd better find out what hobbits were like. But that's only the beginning.

(*Biography*, p. 172; see also *Letters*, p. 215)

Beginning it was, but it was also for Tolkien, as for Bilbo finding the ring on the tunnel-floor in chapter 5 of *The Hobbit*, 'a turning-point in his career'. We know now that Middle-earth, in a sense, already existed in Tolkien's mind, for since at least 1914 he had been writing the elvish and human legends which would appear, many years later and after his death, as the published *Silmarillion* and *Book of Lost Tales*. But Middle-earth would never have caught the public attention without hobbits.

So, what *are* hobbits? And how did Tolkien come to write the seminal sentence in that momentary gap when an alert concentration on tedium suddenly slackened, and allowed, one might imagine, something long repressed or long incubating to break free? Where did hobbits come from, as an idea?

To this last question there are several answers, of increasing levels of interest and complexity. Perhaps the simplest and least satisfying one is gained by looking the word 'hobbit' up in the dictionary – specifically, in the *Oxford English Dictionary*, a gigantic collective project more than a century old, which Tolkien had himself worked for and contributed to in his youth, but which he perhaps as a result continually disagreed with and even went out of his way (in *Farmer Giles of Ham*) to mock. The second edition of the *OED*, published in 1989, says only, 'In the tales of J.R.R. Tolkien . . . one of an imaginary people, a small variety of the human race, that gave themselves this name' (etc.), which gets us no further. However Robert Burchfield, former chief editor of the *OED*, reported with some pride in the *Times* for 31st May

2

1979 that hobbits had at last been run to earth. The word did exist before Tolkien. It is found, once, in a publication called *The Denham Tracts*, a series of pamphlets and jottings on folklore collected by Michael Denham, a Yorkshire tradesman, in the 1840s and 1850s, and re-edited by James Hardy for the Folklore Society in the 1890s. 'Hobbits' appear in Volume 2 (1895). There they come, by my count, 154th in a list of 197 kinds of supernatural creature which includes, with a certain amount of repetition, barguests, breaknecks, hobhoulards, melch-dicks, tutgots, swaithes, cauld-lads, lubberkins, mawkins, nick-nevins, and much, much else, along with the relatively routine boggarts, hob-thrusts, hobgoblins, and so on. No further mention is made of hobbits, and Hardy's index says of them, as of almost all the items in the list, only 'A class of spirits'. Tolkien's hobbits, of course, are anything but 'spirits'. They are almost pig-headedly earthbound, with (as Tolkien wrote in his very earliest account of them, on page 2 of *The Hobbit*):

> little or no magic about them, except the ordinary everyday sort which helps them to disappear quietly and quickly when large, stupid folk like you and me come blundering along making a noise like elephants which they can hear a mile off.

It is *possible* that Tolkien read *The Denham Tracts*, picked up the word 'hobbit', and then forgot all about it till the moment of the blank exam script, but whatever the *Times* may say, the single-word appearance can hardly be called his source, still less his 'inspiration'. Philologists love words, true, but they also know what they are: the word is not the thing.

Not on its own, anyway, for we should remember that Tolkien was keenly interested in words, and names, and their origins, and knew more about some kinds of them than anyone alive (see further pp. 57–9 and 82–6 below). This thought leads to an only slightly more productive theory about hobbits, which is that they sound rather like and therefore might have something to do with rabbits. Shortly after *The Hobbit* came out, on 16th January 1938, the *Observer* printed a letter from an unknown correspondent suggesting some evidently unconvincing connections between hobbits and other real or rumoured furry creatures. Tolkien replied to the correspondent (he did not mean the *Observer* to print his letter, but they did), good-humouredly denying the suggestions, and rejecting both furriness and rabbits:

> my hobbit . . . was not furry, except about the feet. Nor indeed was he like a rabbit . . . Calling him 'a nassty little rabbit' was a piece of vulgar trollery, just as 'descendant of rats' was a piece of dwarfish malice.
>
> (*Letters*, p. 30)

One has to say, however, that it was not just the trolls. The eagle carrying Bilbo in chapter 7 tells him, 'You need not be frightened like a rabbit, even if you look rather like one'. In the previous chapter Bilbo had himself started 'to think of being torn up for supper like a rabbit', and at the end of his stay in Beorn's house Beorn picks him up, pokes his waistcoat disrespectfully, and remarks, 'Little bunny is getting nice and fat again on bread and honey'. Thorin shakes him 'like a rabbit' in chapter 17. The opinion that hobbits are like rabbits is, it seems, pretty widespread among those who meet them. Just the same one can see why Tolkien so firmly rejected the connection. He did not want hobbits, and Bilbo in particular, to be equated with bunnies, or even coneys (another word for 'rabbits' which Bilbo uses): small, fluffy, harmless, irretrievably childish, never rising above the status of pet. The word 'rabbit' was probably professionally interesting to Tolkien, and may have had something to do with the relationship between hobbits and the other races of Middle-earth, for reasons to be explained later on. But whatever else might be said about them, hobbits had to be allowed to be people: not spirits, not animals, but people.

What kind of person? Here one can learn a lot, as might be expected, from the very careful and unexpectedly suggestive presentation of Bilbo right at the start of *The Hobbit* begins, indeed, with the famous sentence of inspiration, the sentence from the subconscious: 'In a hole in the ground there lived a hobbit.' But we are immediately told that this, on its own, would be totally misleading. Creatures that live in holes in the ground ought to be animals – rabbits, moles, snakes, gophers, badgers – and 'hole' conveys a poor impression as a place to live. 'Don't call my palace a nasty hole!' says Thorin much later, in chapter 13. 'You wait till it has been cleaned and redecorated!' Bilbo's hole, however, needs neither cleaning nor redecorating, for the description goes on, firmly and rhythmically negating all the suggestions of the sentence before it:

> Not a nasty, dirty, wet hole, filled with the ends of worms
> and an oozy smell, nor yet a dry, bare, sandy hole with
> nothing in it to sit down on or to eat: it was a hobbit-hole,
> and that means comfort.

It is in fact, in everything except being underground (and in there being no servants), the home of a member of the Victorian upper-middle class of Tolkien's nineteenth-century youth, full of studies, parlours, cellars, pantries, wardrobes, and all the rest.

Bilbo himself is furthermore fairly easy to place both socially and even chronologically. If one did not have the rest of the book to go on, one would have to place him, on internal evidence, from a time after the discovery of America, for he smokes a pipe, and indeed the last words of the whole book are 'tobacco-jar' ('tobacco' is not recorded in English by the *OED* till 1588). But one could be more precise than that, for when Bilbo wishes to discourage Gandalf he takes out 'his morning letters', which are clearly routinely delivered early every day. Bilbo must live, then, after the introduction of a postal service – our familiar system dates, in England, from 1837. In a more indirect way Bilbo might also be thought to date from a time after railway-engines, for though it is the narrator's term not his own, when his nerve finally breaks he shrieks 'like the whistle of an engine coming out of a tunnel' (the first freight-and-passenger steam railway in England opened in 1825, the first railway tunnel dating from five years later).

All this of course turns out to be completely wrong, and we are told point-blank that the story is set 'long ago in the quiet of the world, when there was less noise and more green'. Tolkien, however, did not forget any of the points raised above, and would later go to some lengths to explain them away, or blur them. But the fact is that hobbits are, and always remain, highly *anachronistic* in the ancient world of Middle-earth. That indeed is their main function, for one might note that by their anachronism they engage a problem faced and solved in not dissimilar ways by several writers of historical novels. In setting a work in some distant time, an author may well find that the gap between that time and the reader's modern awareness is too wide to be easily bridged; and accordingly a figure essentially modern in attitudes and sentiment is imported into the historical world, to guide the reader's reactions, to help the reader feel 'what it would be like' to be there. An obvious example comes from the Hornblower novels of C.S. Forester, which began to be published at exactly the same time as *The Hobbit*. In them, as all readers of them will remember, the hard-headed and hard-hearted Bush stands for Nelsonian normality, firmly contrasted with the more intelligent, more squeamish, and much more twentieth-century figure of Hornblower, with his horror of flogging, belief in cold showers and cleanliness, and dangerously democratic notions. Bilbo, even more than his successor-hobbits from *The Lord of the Rings*, takes up this role as 'reflector'. His failings are those which the child reader, and indeed the adult reader, would have if transported magically to Middle-earth. He is 'used to having [his meat] delivered by the butcher all ready to cook', has no idea how to 'hoot twice like a barn-owl and once like a screech-owl', and has to cover up his inability to understand anything of bird language, whether 'quick and difficult' or not. He is a modern person, or at least a twentieth-century person, who seems again and again to be out of place in the archaic and heroic world into which he is drawn, or thrust, by Gandalf.

On the other hand, Bilbo is solidly placed in hobbit society, which requires no explanation at all (at least for the reader of 1937). Once his 'hole' has been dealt with, and any incorrect suggestions the word may have created have been explained away, the first thing we are told about Bilbo is his social standing: and this is unusually precise. Thus Bilbo is 'well-to-do', but not necessarily 'rich'; most of his paternal relations are rich, but not as rich as his maternal ones. The *OED*, here an excellent guide, as to most Victorian or Edwardian usage, defines 'well-to-do' as 'Possessed of a competency; in easy circumstances', by which it means above all, not having to work. 'Rich' by contrast has several meanings, being an old word, but the relevant one is 'Having large possessions or abundant means' – abundant as opposed to competent. Bilbo then has enough and a bit over, but not more than that. What he and his family do have without qualification, however, is 'respectability', which in English society had and still has no correlation whatever with wealth. It is perfectly possible, indeed normal, to be a respectable member of the working classes, and just as normal to be a member of the upper classes with no respectability whatsoever. The *OED* defines 'respectable' carefully as 'Of good or fair social standing, and having the moral qualities naturally appropriate to this': note the words 'or fair', with which Tolkien would have agreed (there is no doubt later on that the Gamgee family is respectable, and capable of major social mobility, but without even a 'competency' to start with); and also the undefined and unconsidered 'naturally appropriate', which Tolkien would probably have regarded as yet another example of the dictionary editors' incurable smugness. Bilbo is in short middle-middle to upper-middle class. Though there is one counter-indication to this, which is that his name is Baggins.

Baggins is incipiently vulgar. One of the trolls, who are very vulgar, as Tolkien said (see above), is called Huggins, indeed Bill Huggins, not so far from Bilbo Baggins. Huggins meanwhile – I repeat that Tolkien knew a great deal about names – is a diminutive form of a personal name (Hugh, Hugo), like the common surnames Watkins, Jenkins, Dickens, and so on. Baggins, however, isn't, though it is a common word in two senses. It is 'common' in not being standard, therefore (in post-medieval England, but not earlier) vulgar, low-class, dialectal; and it was in common (i.e. general) use across the whole of Northern England to mean the food a labourer takes with him when he goes off to work, or anything eaten between meals, but especially, says the *OED*, afternoon tea 'in a substantial form'. Tolkien certainly knew this, and knew also that the *OED* had tidied the word up from 'baggins' (which is what people really say) to 'bagging' (which is hyper-correct), for the word is cited and defined in the *New Glossary of the Dialect of the Huddersfield District* to which he had written an appreciative prologue in 1928 – Tolkien was not a Northerner, but he remained all his life grateful and even 'devoted' to the University of Leeds (see *Letters*, p. 305), and appreciative of Northern dialect. *The Hobbit* indeed ends with a joke derived from the *Glossary* just mentioned, for in the Huddersfield dialect the word 'okshen' meant not 'auction' but 'mess'. Walter Haigh, who compiled the *Glossary*, records the disapproving sentence, used seemingly by one woman of another, 'Shu'z nout but e slut; er ees [her house] ez e feer okshen [a right mess]'. And when Bilbo returns home, what he finds is an 'okshen' in both senses, mess and auction at once.

To return to Bilbo Baggins, though, he is fond of all meals, as we soon learn, but most especially his tea. The 'Unexpected Party' of chapter 1 is definitely a tea-party, and undeniably substantial. This makes another anachronistic point about Bilbo, and about hobbits in general, which is that they are very specifically English. Tolkien was to rub this point in very firmly indeed in the 'Prologue' to *The Fellowship of the Ring*, in which he makes the whole history of the Shire correspond point for point with the history of early England. But it is clear enough from Bilbo's first encounter with Gandalf. Not to make too much of it, Bilbo is something of a snob: not a terrible one, for he is prepared to offer a pipe to passing strangers, but certainly liable to draw a line between 'his sort' and other sorts. At several points he displays the social exclusiveness which has so often annoyed visitors to England. He dismisses the whole idea of 'adventures' with 'I can't think what anybody sees in them', and then tries to get rid of Gandalf, whom he has decided is 'not quite his sort' by ignoring him. He goes on, with entirely insincere politeness, to try to send Gandalf away by repeating 'Good morning!' as a parting not a greeting, to try 'thank you!' in the same spirit, twice (it means, when said in clipped English tones, '*no* thank you'), and eventually to invite him to tea – but not now. It is obvious that much of what Bilbo says is socially coded to mean its opposite, as when a few pages later he says to the dwarves, 'in his politest unpressing tones', 'I suppose you will all stay to supper?' (which means, to those who know the code, 'you have overstayed your welcome, go away').

None of this is unfamiliar at all to the English reader, and of course it is comic to find Gandalf repeatedly ignoring the social code, and acting, as only someone foreign to it would, as if Bilbo meant what he said by phrases like 'I beg your pardon'. There is in fact a word which sums Bilbo up, often used of the English middle-class to which he so obviously belongs: 'bourgeois'. This is not an English word but a French one, and Tolkien does not use it – he regretted, again for professional reasons, the medieval take-over of the English language by Norman French, and always tried to reverse it as far as he could. But he may well have been thinking of just that word, as is indicated by a couple of running private jokes. Later on, in *The Lord of the Rings*, it will be disclosed that the road Bilbo's hole is on is called Bag End: very appropriate for someone called Baggins, perhaps, but an odd name for a road. And yet in a sense a very familiar one. As part of the ongoing and French-oriented snobbery of English society in Tolkien's day (and later), municipal councils were (and still are) in the habit of indicating a street with no outlet as a 'cul-de-sac'. This is French, of course, for 'bag end', though the French actually call such a thing an *impasse*, while the native English is 'dead end'. 'Cul-de-sac' is a silly phrase, and it is to the Baggins family's credit that they will not use it. The Tolkien family's too, for his Aunt Jane Neave's house was down a lane with no exit, also defiantly called 'Bag End' (see *Biography*, p. 106). It is a very bad mark for the socially aspiring branch of the Baggins family that they have tried to Frenchify themselves and disguise their origins: they call themselves the Sackville-Bagginses, as if they came from a *ville* (or villa?) in a *cul-de-sac(k)* (Bag End). They, then, are really *bourgeois*. Bilbo is just heading that way.

Gandalf means, however, to turn him back, and that is why he makes him a burglar. 'Burglar' is another odd word, and English speakers who use it tend to assume that the -ar on the end is the same as -er. Accordingly, just as a worker is someone who works, so a burglar must be someone who burgles. But this derivation is false, and exemplifies two things Tolkien yet again knew a great deal about, 'back formation' and 'folk etymology'. The root of 'burglar' is in fact the same as that of 'bourgeois', Old English (and probably Old Frankish too) *burh*, 'borough, town, fort, stockaded mansion'. A *burgulator*, as the *OED* points out, is someone who breaks into mansions, a *bourgeois* is someone who lives in one. They are connected opposites, like Sackvilles and Bagginses. Gandalf means to move Bilbo from the one side, the snobbish side, to the other.

In doing this Bilbo will not become less English, but more so. We should note, in view of the bad press which 'Englishness' has had for most of the twentieth century, that Tolkien was quick to point out some of Bilbo's native virtues, in terms quite similar to those of George Orwell, another contemporary of Tolkien's, and another example of English 'self-fashioning' (for Orwell's real name was Blair, which he abandoned because he thought it sounded Scottish, just as Tolkien, aware that his own name was originally German, tended to identify himself with his mother's Worcestershire family name of Suffield, *Letters*, p. 218). The narrator comments, once Bilbo has recognized Gandalf and responded with genuine excitement and interest, 'You will notice already that Mr. Baggins was not as prosy as he liked to believe, also that he was very fond of flowers'. Hobbits, then, like the English middle class to which they clearly belong, may aspire to be bourgeois and boring, but it is not natural to them. Tolkien indeed had nothing against middle-class Englishmen, for he was one himself: and, unlike so many of the English-speaking writers of his time, Lawrence, Forster, Woolf, Joyce, he did not feel in any way alienated, nor have any urge to reinvent himself as working-class, non-English, in internal exile, or any other glamorous pose. It is one reason why he has never found any favour with the determinedly cosmopolitan British *intelligentsia* (to use another foreign term).

Bilbo is then defined from the start by time, class, and culture. He is English; middle class; and roughly Victorian to Edwardian. Hobbits in general will prove to be all these things even more definitely than Bilbo, except that some of them will be working class (the Gamgees), though none quite reach the upper class, not even the Tooks and Brandybucks. But he and they are also repeatedly marked as anachronisms in the world they inhabit. On the surface at least – the issue is explored all the way through *The Hobbit* and *The Lord of the Rings* – they do not fit at all into Middle-earth, the world of dwarves and elves, wizards and dragons, trolls and goblins, Beorn and Smaug and Gollum.

The world of fairy-tale

This world is not, in origin, Tolkien's invention: though it is perhaps his major achievement to have opened it up for the contemporary imagination. In 1937 (though not now) the world and its personnel were best known from a relatively small body of stories taken from an again relatively small corpus of classic European fairy-tale collections, those of the Grimm brothers in Germany, of Asbjørnsen and Moe in Norway, Perrault in France, or Joseph Jacobs in England, together with literary imitations like those of H.C. Andersen in Denmark, and literary collections like the 'colour' Fairy Books of Andrew Lang; and from the many Victorian 'myth and legend' handbooks which drew on them. These tales made concepts like 'dwarf' or 'elf' or 'troll' familiar to most people from early childhood. Dwarves, for instance, figure prominently in 'Snow White', and share some of the characteristics of Thorin's people, like their mining profession and their fascination with wealth. Trolls were not so well-known in English (the word is a Scandinavian one), but just the same have entered English consciousness through 'The Three Billy-Goats Gruff', a tale recorded by the Norwegians Asbjørnsen and Moe. Elves appear in the tale of 'The Little Elves and the Shoemaker', and goblins in the literary fairy-tale imitations of George MacDonald. Few children grow up without encountering some of these stories, and others like them.

These traditional fairy-tales, however, have severe limitations in at least two ways. One is that they are detached from each other. There may be a vague sense that they all take place in something like the same world, a dimly-perceived far past which, as Bilbo says of Gandalf's stories, is all about 'dragons and goblins and giants and the rescue of princesses and the unexpected luck of widows' sons'. But this world is connected to no known history or geography, and furthermore there is no connection between any of the tales themselves. They cannot, then, be developed. They stimulate the imagination, but do not entirely satisfy it – not, at least, in the way that modern readers expect, with a full plot and developed characters and, perhaps most of all, a map.

And there is another problem with fairy-tales which Tolkien sensed very keenly. This is that from their very beginning, from the time, that is, when scholars began to take an interest in them and collect them, they seemed already to be in a sense in ruins. The Grimm brothers, in the nineteenth century, quite certainly had as a main motive for making their collection of *Haus- und Kindermärchen* the wish to do a kind of literary rescue archaeology. They were convinced that the tales they collected, brief as they were and deep sunk in the social and literary scale, still preserved fractions of some older belief, native to Germany but eventually suppressed by foreign missionaries, foreign literacy, and Christianity. Jacob Grimm, the elder brother, indeed tried to fit the pieces together, or at least collect as many as he could, in his extensive work *Deutsche Mythologie*, or 'Teutonic Mythology'. The attempt has since then been generally ignored or derided, but there were some true observations behind it. One was that in some cases, like 'dwarf' (see pp. xiv–xv above), all Germanic languages had preserved the same word; though they had clearly not borrowed it from each other, because the word had always changed as the languages had changed, over the millennia. Accordingly English speakers said 'dwarf', Germans *Zwerg*, and Icelanders *dvergr*. What this seemed to indicate was that the word was very old, much older than the fairy-tales in which it was preserved. But it must have been used in fairy-tales all along. What could those old tales have been like, before the whole mythology had been downgraded to children and their nursemaids?

Strong confirmation of this theory furthermore came from the rediscovery, during the eighteenth and nineteenth centuries, of fragments of the old adult and aristocratic literature of Northern

Europe. It should be noted that this had been totally lost and forgotten for many centuries – Shakespeare, for instance (though he clearly knew something about fairy-tales, more than he was prepared to show) can have known nothing about the higher literature that lay behind them. The one surviving copy of the Old English epic *Beowulf*, with its strong interest in monsters, elves and orcs included, lay as far as we can tell unread and almost unnoticed from the Norman Conquest in 1066 till its eventual publication in Copenhagen in 1815. The Old Norse poems of the *Elder Edda* likewise lay unknown and for the most part in one manuscript in an Icelandic farmhouse, till they were rediscovered and slowly and patchily republished by scholars including the Grimms. The Middle English poem *Sir Gawain and the Green Knight*, with its very similar interest in elves and ettins, was hardly known and certainly not a part of university syllabuses till it was edited by Tolkien himself and his junior Leeds colleague E.V. Gordon in 1925. To those, however, who *did* read these poems and their many badly-preserved analogues, there came a feeling that their authors had indeed known something, something consistent with each other and with the much later fairy-tales of modern times: and that you might just possibly be able to work out what it was. This is the philological activity of 'reconstruction', as discussed in the Foreword above, p. xv.

In two ways, then, fairy-tale and its ancestors provoked the imagination, suggested a wider world which they then did not explore. You could work back from the dwarves of 'Snow White'; or you could work out from the dwarves of *Ruodlieb* (a poem written in Latin by a German poet of the twelfth century) or of the *Elder Edda* (a collection of poems written in Old Norse, some of them probably even older than *Ruodlieb*). This is what Tolkien was doing: as is proved, for instance, by his stubborn insistence on writing, and making the printers print the word 'dwarves', even though (as he says in his opening note to *The Hobbit*) 'In English the only correct plural of *dwarf* is *dwarfs*'. If that is the only correct form, why use an incorrect one? Because the -ves ending is a sign of the word's antiquity, and so its authenticity. Even in modern English, old words ending in -f make their plural with -ves, as long as they have remained in constant use: so hoof/hooves, life/lives, sheaf/sheaves, loaf/loaves. Dwarf/dwarves might have developed the same way, but clearly fell out of general use, and so was assimilated (probably by literates, schoolteachers and printers) to the simpler pattern of tiff(s), rebuff(s), and so on. Tolkien meant to turn back this particular clock. The Grimms had done exactly the same with their insistence that the German plural for 'elf' ought to be *Elben*, not *Elfen* (a form borrowed late on from English, itself by that time historically mistaken). Tolkien furthermore quite clearly had in mind from the start of *The Hobbit* a poem from the *Elder Edda*. It gave him all the names of 'Thorin and Company'.

There are, one might note, surprisingly few names in *The Hobbit*, certainly by comparison with *The Lord of the Rings*. Most natural features have names which are just common nouns and adjectives with capital letters, like The Hill, The Water, Dale, the Long Lake, the River Running, the Lonely Mountain, Ravenhill and indeed The Carrock. To Bilbo's timid question about the meaning of the last Gandalf replies crushingly that '[Beorn] called it the Carrock, because carrock is his word for it. He calls things like that carrocks, and this one is *the* Carrock because it is the only one near his home'. In addition to this we have a few hobbit-names (Baggins and Took, Hobbiton, and the auctioneers Grubb, Grubb, and Burrowes); rather more names from Tolkien's already developed but here only hinted-at elvish mythology (Elrond, Gondolin, Girion, Bladorthin, Dorwinion, and more doubtfully Orcrist and Glamdring); and a few incidentals (Radagast, Bolg and Azog the goblins, Carc and Roäc the ravens, Bard). But when it comes to dwarf-names, Tolkien gives full measure.

He found them in the poem *Völuspá*, 'The Sybil's Vision', one section of which is called the *Dvergatal*, 'the Tally of the Dwarves'.

In the original Old Norse, this contains rather more than sixty names, mostly strung together as a simple rhythmic list, repeated in slightly different form in Snorri Sturluson's thirteenth-century guide to Norse mythology, the *Skaldskaparmál*, the 'Treatise on Skald-ship', or one might say, 'Art of Poetry'. Part of Snorri's version goes as follows, and one can see immediately the connection with Tolkien:

> Nár, Náinn, Nípingr, Dáinn,
> Bífur, Báfur, Bömbur, Nóri,
> Órinn, Ónarr, Óinn, Miöðvitnir,
> Vigr og Gandálfr, Vindálfr, Þorinn,
> Fíli, Kíli, Fundinn, Váli,
> Þrór, Þróinn, Þettr, Litr, Vitr . . .

Eight of the thirteen dwarf-names of Tolkien's Thorin and Company are here, along with the name of Thorin's relative Dain, his grandfather Thror, and something close to his father Thráin. Four of the other five (Dwalin, Gloin, Dori, Ori) are not far away, as are Durin, in both *The Hobbit* and *Völuspá* the dwarves' legendary ancestor, and Thorin's nickname Oakenshield, or Eikinskjaldi. Only Balin – a famous name in Arthurian story, though that is perhaps a coincidence – is not in Snorri's list.

However Tolkien did not just copy the 'Tally of the Dwarves', or quarry it for names. He must rather have looked at it, refused to see it, as most scholars do, as a meaningless or no longer comprehensible rigmarole, and instead asked himself a string of questions about it. What, for instance, is 'Gandálfr' doing in the list, when the second element is quite clearly *álfr*, 'elf', a creature in all tradition quite distinct from a dwarf? And why is 'Eikinskjaldi' there, when unlike the others it does not seem to be a possible name, but looks like a nickname, 'Oakenshield'? In Tolkien of course it *is* a nickname, the origin of which is eventually given in Appendix A (III) of *The Lord of the Rings*. As for Gandálfr, or

Gandalf, Tolkien seems to have worked out a more complex explanation. In early drafts of *The Hobbit* Gandalf was the name given to the chief dwarf, while in the first edition what Bilbo sees that first morning is just 'a little old man'. Even in the first edition, however, the little old man's staff soon comes into the story, while by the third edition – Tolkien made significant changes in both the second and third editions, 1951 and 1966, some of them discussed later on – Gandalf has become 'an old man *with a staff*' (my emphasis). This seems highly suitable. Even now the 'magic wand' is the common property of the stage-magician, while in all popular and learned literary tradition, from Shakespeare's Prospero to Milton's Comus or Terry Pratchett's Discworld, the staff is the distinguishing mark of the wizard. It looks as if Tolkien sooner or later interpreted the first element of 'Gandálfr', quite plausibly, as 'wand' or 'staff', while the second element, as said above, obviously means 'elf'. Now Gandalf in Tolkien is definitely not an elf, but then it turns out that he is not just an 'old man' either; one can see that to those who knew no better (people like Éomer in *The Lord of the Rings* much later on) he might well seem distinctly 'elvish'. Tolkien seems to have concluded at some point that 'Gandálfr' meant 'staff-elf', and that this must be a name for a wizard. And yet the name is there in the *Dvergatal*, so that the wizard must in some way have been mixed up with dwarves. Could it be that the reason the *Dvergatal* had been preserved was that it was the last fading record of something that once had happened, some great event in a non-human mythology, an *Odyssey* of the dwarves? This is, anyway, what Tolkien makes of it. *The Hobbit*, one might say, is the story that lies behind and makes sense of the *Dvergatal*, and much more indirectly gives a kind of context even to 'Snow White' and the half-ruined fairy-tales of the brothers Grimm.

The author's voice

The two sides of *The Hobbit* are, then, fairly clear: on the one side there is modern middle-class English Bilbo, on the other the archaic world which lies behind both vulgar folk-tale and its aristocratic, indeed heroic ancestors. The former is represented by clocks and fussiness – Bilbo gasping out, 'I didn't get your note till after 10.45 to be precise', and feeling he cannot leave home without a pocket-handkerchief. The latter is created by poetry and the Misty Mountains and Bilbo feeling how grand it would be to 'wear a sword instead of a walking-stick'. Naturally the two sides are going to clash, and much of *The Hobbit* is about the clash of styles, attitudes, behaviour patterns – though in the end one might conclude that they are not as far apart as they first seemed, and that Bilbo has just as much right to the archaic world and its treasures as Thorin or Bard. However the pressing problem for Tolkien was perhaps not to introduce the archaic world – much of which, as has been said above, has long been familiar at least in its personnel even to child readers – as to give it intellectual coherence, to make the reader feel that it had a sort of existence outside the immediate narrative. Tolkien solved this problem, in *The Hobbit*, if quite differently in *The Lord of the Rings*, by flexible and intrusive use of the authorial voice.

The general strategy is shown several times in the first few pages. At the start of paragraph four Tolkien imagines a question from a reader, 'what is a hobbit?', and replies as if hobbits are not unknown but may have escaped some readers' attention: 'I suppose hobbits need *some* description *nowadays*' (my emphasis both times). As soon as that parenthesis is over, we are told that Bilbo's mother was 'the *famous* Belladonna Took' (my emphasis again), the implication again being that the author is only selecting from a body of pre-existing information. Her distinction is partly explained by the theory that 'one of the Took ancestors must have taken a fairy wife', immediately corrected – in the 1966 edition, previous editions being slightly different – by 'That was, of course, absurd'. This time the word 'absurd' implies that there are well-known ways of judging such statements, so well-known that the author has no need to give them, while the 'of course' assumes that the reader must know these too. In every case the suggestion is that there is story outside the story, so to speak, a whole wider world of which one is seeing only some small fraction. The point is made totally explicit on page 3, when the narrator interjects 'Gandalf! If you had heard only a quarter of what I have heard about him, you would be prepared for any sort of remarkable tale'.

All these devices are furthermore repeated, sometimes again and again – in his essay on 'Some of Tolkien's Narrators' in the very recent collection *Tolkien's 'Legendarium'* (for which see the final 'List of References'), Paul Edmund Thomas lists some 45 cases of direct address by the narrator of *The Hobbit*, and this does not include some of the types of interjection discussed here. 'The famous Belladonna Took' is echoed by 'the great Thorin Oakenshield himself' (the reasons for his fame and nickname given only in a footnote in Appendix A (III) of *The Lord of the Rings*, seventeen years and a thousand pages later). Even in the first edition Tolkien used the 'of course' trick at least three more times in exactly the same way as with the 'fairy wife' theory: 'They were elves of course', in chapter 3, 'The feasting people were Wood-elves of course' (chapter 8), and 'That of course is the way to talk to dragons' (chapter 12). Very similar is the trick of suddenly producing a piece of totally unexpected and unpredictable information from the heart of the fairy-tale world, and pretending it is common knowledge. The most dramatic example is the climax of the troll episode, when the light comes over the hill and the birds twitter, and the trolls turn to stone: 'for trolls, *as you probably know*, must be underground before dawn, or they go back to the stuff of the mountains they are made of'. The idea is indeed an ancient one, with Odin playing the same trick as Gandalf (though on a dwarf) in the Old Norse poem *Alvíssmál*. But in context it is totally unexpected, though Tolkien had prepared for it with earlier direct addresses to the reader, once again invoking prior knowledge: 'Yes, I am afraid trolls do behave like that, even those with only one head each . . . it is nearly always worth while, if you can manage it . . . Trolls' purses are the mischief, and this was no exception'. There is a kind of unfairness in it, for the author naturally knows everything and the reader nothing about the world being introduced, but the voice assumes a kind of complicity; and every time another piece of the picture is being filled in, another part of the mental map disclosed. By the end of *The Hobbit* – and this was one of the reasons for the immediate demand for a sequel – a detailed and consistent picture of the fairy-tale world, and of many of its inhabitants, had been generated. Tolkien had to set the scene, indeed to guarantee that there was a scene to set, before the story could be allowed to unroll.

The story itself is highly episodic, and so not easy to summarize. Briefly, one may say that the book's nineteen chapters divide approximately half and half into the adventures which Bilbo and the dwarves have *before* they reach the Lonely Mountain and the lair of Smaug the dragon; and the complexities surrounding the gaining, guarding and sharing of the dragon's treasure once the Lonely Mountain has been reached. Chapters tend to come in threes, with numbers 1 to 3 getting the company as far as the Misty Mountains, where they are captured by the goblins; 4 to 6 dealing with the crossing of the mountains, including Bilbo winning the magic ring of invisibility; and 7 to 9 set in Mirkwood, where Bilbo uses his ring twice to rescue the dwarves first from the giant spiders, and second from the Wood-elves' prison. Chapters 12 to 14 deal with Bilbo's first two attempts to 'burgle' Smaug, the dragon's attempted revenge and final death at the hands of Bard the Bowman; and chapters 15 to 17 with the quarrels over the treasure, between dwarves, elves, men and eventually goblins. The last two chapters are an evident coda, returning Bilbo to his home; while the two central chapters 10 and 11 mark a kind of transition, as Bilbo emerges for a short while from an entirely archaic and romantic world to a world once more dominated by human beings, humdrum ideas of 'business' and the Master of Laketown, even more of a bourgeois than Bilbo.

None of these divisions, of course, is vital, and it is quite likely that Tolkien did not plan them or pay any attention to them. They do show, however, how Tolkien fed in the fairy-tale elements one at a time, introducing them separately for many chapters before making much attempt to combine them, so that they go, in order of chapters: dwarves (and a wizard); trolls; elves; goblins; Gollum; wargs and eagles; Beorn; wood-elves and spiders (so chapters 1 to 8), with after that only one entirely novel figure introduced – Smaug in chapter 12 – and the interaction of all the creatures previously mentioned, apart from Gollum and the trolls, in the negotiations over the treasure and the Battle of the Five Armies. The other aspect of this one-at-a-time presentation, though, is the steady rise of Bilbo's status, and the increasing evenness of the confrontation between the modern values he represents and the ancient ones he encounters.

The contest for authority

At the start of the book Bilbo, as befits his bourgeois status and anachronistic nature, is helpless and, if not contemptible, at least open to contempt from those around him. Thorin's casually gloomy speech, which takes the violent death of some or all of his company as a matter of course, frightens him into a screaming fit which even Gandalf has difficulty explaining away. Glóin's 'He looks more like a grocer than a burglar' might not be much of a condemnation in other circumstances, but in the heroic world it is. No one in any medieval epic or Norse saga could possibly behave like Bilbo. The cook who begs for his life in the Eddic poem *Atlamál* (said to have been written in Greenland) is regarded as nothing but a figure of fun; the old man who bursts into tears in *The Saga of Hrafnkel Priest of Frey* (edited a few years before *The Hobbit* by Tolkien's former colleague, E.V. Gordon) is viewed so scornfully that the place where he cried is still, the saga-author says, called *Grátsmýrr*, 'Greeting-moor' ('greet' remains the northern English and Scottish dialect word for 'weep'). It is true that Bilbo recovers himself and gets back on his dignity, abetted by Gandalf, but he still has to be apologized for: 'He was only a little hobbit you must remember'.

He does only slightly better in the scene with the trolls, for though he does try to intervene in the fight – 'Bilbo did his best' – it is so ineffectual that no one notices. He does feel a kind of pressure to conform to the expectations of the fairy-tale world (which includes stories like the Grimms's 'The Brave Little Tailor', rather similar to this scene, and Asbjørnsen and Moe's 'The Master-Thief', which is what Bilbo would like to be), for he tries to pick the troll's pocket, because 'somehow he could not go straight back to Thorin and Company emptyhanded'. But his complete ignorance about trolls' purses makes that a failure, while the one physical ability we do hear about, that 'hobbits can move quietly in woods, absolutely quietly', is counterbalanced by his inability to do another thing the dwarves take for granted, 'hoot twice like a barn-owl and once like a screech-owl'. Bilbo does not show himself up this time, and he does find the trolls' key, but he remains comically out of place.

The pattern is repeated in chapter 4, where Bilbo has to be carried in the escape from the goblins, and where both he and Bombur agree that he is out of place quite literally, with their antithetical 'why did I ever leave my hobbit-hole! . . . why did I ever bring a wretched little hobbit on a treasure-hunt!' However, just as it was conceded that hobbits could at least move quietly, so here it is conceded that Bilbo does at least do something useful, in waking up and letting out the yell that warns Gandalf. But so far it is fair to say that he has done nothing that might seem impossible for a child-reader imagining a similar situation.

This changes with Bilbo's discovery of the ring, 'a turning point in his career, but he did not know it', as Tolkien notes (with a certain irony, since it was a turning point for Tolkien too, though in 1937 he was even less aware of this than was Bilbo). After he has found it, Bilbo continues to think of his hobbit-hole and 'himself frying bacon and eggs in his own kitchen', a characteristically modern and characteristically English menu, while he also, with yet another anachronism, gropes for matches for his pipe (friction matches were invented in 1827). But then he remembers his sword, draws it, realizes 'it is an elvish blade' like Orcrist and Glamdring, and feels comforted. 'It was rather splendid to be wearing a blade made in Gondolin for the goblin-wars of which so many songs had sung', says the narrator, and though this romantic sentiment is immediately qualified by a practical one – 'and also he had noticed that such weapons made a great impression on goblins that came upon them suddenly' – it marks perhaps the first stage in Bilbo's winning a place in the world of fairy-tale. The narrator follows this up by distancing Bilbo a little from modern times and from the child-reader. He was in a tight place, yes, 'But you must remember it was not quite so tight for him as it would have been for me or for you'. Hobbits, after all, 'are not quite like ordinary people'. They do live underground; they move quietly (which we knew already); recover quickly; and most of all 'they have a fund of wisdom and wise sayings that men have mostly never heard or have forgotten long ago'.

Bilbo's riddle-exchange with Gollum actually falls mostly into the latter category, of things forgotten, for the whole idea of testing by riddles, and some of the actual riddles, come from the ancient and aristocratic literature of the Northern world rediscovered in the nineteenth century by Tolkien's professional predecessors. Gollum asks five riddles and Bilbo four – his fifth being the non-riddle 'What have I got in my pocket?' – and of these nine, several have definite and ancient sources. They probably all have sources – Tolkien's 1938 letter in the *Observer* had teasingly said as much, see *Letters* (p. 32), and Douglas Anderson's *Annotated Hobbit* of 1988 identifies as many as possible – but Gollum's riddles, unlike Bilbo's, tend to be ancient ones. Thus his last riddle, delivered when he thinks 'the time had come to ask something hard and horrible', derives from a poem in Old English, the riddle-game, or more precisely the wisdom-testing exchange, between Solomon and Saturn. In this, Saturn, who represents heathen knowledge, asks Solomon, 'What is it that . . . goes on inexorably, beats at foundations, causes tears of sorrow . . . into its hands goes hard and soft, small and great?' The answer given in *Solomon and Saturn* is, not 'Time' as in Bilbo's desperate and fluky reply, but 'Old age': 'She fights better than a wolf, she waits longer than a stone, she proves stronger than steel, she bites iron with rust: she does the same to us'. This is a more laboriously dignified version of Gollum's:

> Gnaws iron, bites steel;
> Grinds hard stones to meal;
> Slays king, ruins town,
> And beats high mountain down.

Gollum's 'fish' riddle:

> Alive without breath,
> As cold as death;
> Never thirsty, ever drinking,
> All in mail, never clinking

is echoed by a riddle set in the Old Norse wisdom contest in *The Saga of King Heidrek the Wise* (to be edited many years later by Tolkien's son Christopher), and has a further slight analogue in a medieval poem from Worcestershire which Tolkien admired, *Layamon's 'Brut'*: in this dead warriors lying in a river in their mail are seen as strange fish. Gollum's 'dark' riddle – 'something a bit more difficult and more unpleasant' – again has an analogue in *Solomon and Saturn*, though there the answer (and Tolkien was to remember this later) is not 'dark' but 'shadow'. Gollum's riddles, cruel and gloomy, associate him firmly with the ancient world of epic and saga, heroes and sages.

But Bilbo can play the game too; though his riddles are significantly different in their sources and their nature. Three of them, 'teeth', 'eggs', and 'no-legs', come from traditional nursery-rhyme (versions of them are printed in *The Annotated Hobbit*). But where, one might ask, does traditional nursery-rhyme come from? Tolkien had certainly asked himself this question, which relates directly to the point made above about the sources of traditional fairy-tale, long before he began to write *The Hobbit*. In 1923 he had published a long version of the familiar 'man in the moon' nursery-rhyme, 'Why the Man in the Moon Came Down too Soon', eventually reprinted as number six in *The Adventures of Tom Bombadil*. In the same year he published 'The Cat and the Fiddle: A Nursery Rhyme Undone and its Scandalous Secret Unlocked'; this became the hobbit-poem which Frodo sings in the *Prancing Pony* in Bree, in *The Lord of the Rings*, and was also reprinted in *Tom Bombadil*. Later on, his 1949 short story *Farmer Giles of Ham* (which was originally written over about the same period as *The Hobbit*, see Wayne Hammond and Douglas Anderson's *Descriptive Bibliography*, pp. 73–4) is set firmly in the land of nursery-rhyme, of Old King Cole and 'all the king's horses and all the king's men'. Tolkien notes in a letter to Stanley Unwin in 1938 that a friend and Oxford colleague had written a long 'rhymed tale in four books' called *Old King Coel* – Coel, note, not Cole, for there is a 'King Coel' in old Welsh tradition. It may seem surprising that anyone should find nursery-rhymes worth quite so much time and trouble, if it does not quite extend to taking them seriously. But behind all these rewritings and reminiscences lies the philologist's conviction that, just as the children's fairy-tales of elves and dwarves had some long-lost connection with the time when such creatures were material for adults and poets, so modern playground riddles and rhymes were the last descendants of an old tradition. Tolkien had furthermore tried to fill the gap of time, as he often did, in this case by writing a version of the children's 'eggs' riddle in Old English, or Anglo-Saxon. He published this too in 1923, as one of the 'Enigmata Saxonica Nuper Inventa Duo' ('Two Anglo-Saxon Riddles recently Discovered'). Ten lines long, it starts:

> Meolchwitum sind marmanstane
> wagas mine wundrum frætwede . . .

'My walls are adorned wondrously with milk-white marble-stones . . .' This, one might say, is what the modern children's riddle's ancient ancestor must have looked like. It is (see p. xv above) an 'asterisk-riddle'.

When Bilbo replies to Gollum's ancient riddles with modern ones, then, the two contestants are not so very far apart. As Gandalf was to say to Frodo many years later (by which time the concept of Gollum had admittedly changed a good deal), 'They understood one another remarkably well . . . Think of the riddles they both knew, for one thing'. What this suggests, though, is that while Bilbo remains an anachronism, a middle-class English-man in the fairy-tale world, he is indeed 'not quite like ordinary people'. The difference is that he has not quite lost his grip on old tradition. Nor, of course, have all 'ordinary people'. But they have downgraded old tradition to children's tales and children's songs, become ashamed of it, made it into 'folklore'. Bilbo and hobbits are in this respect wiser. Their unforgotten wisdom puts Bilbo for the first time on a level with a creature from the world into which he has ventured.

Bilbo also, after this point, has the ring: in *The Hobbit*, not yet the Ring, but still a potent force to help him gain the grudging respect of the dwarves. He has two other qualities besides. One is luck. The dwarves notice this more than once, with Thorin for instance saying, as he sends Bilbo down the tunnel to the dragon, that Bilbo is 'full of courage and resource far beyond his size, and if I may say so possessed of good luck far exceeding the usual allowance' (chapter 12). Earlier on, after Bilbo had rescued them from the spiders, '[the dwarves] saw that he had some wits, as well as luck and a magic ring – and all three were useful possessions' (chapter 8). This belief that luck is a *possession*, which one can own, and perhaps even give away or pass on, may seem to be characteristically dwarvish, i.e. old-fashioned, pre-modern: it is a commonplace of Norse saga, for instance, where there are many lucky and unlucky cloaks, weapons, and people. But people do not think that way about luck any more. Or do they? In fact, superstitions about the nature of luck remain surprisingly common – they are a repeated sub-theme in Patrick O'Brian's long series of historical novels about the nineteenth century, though one should note that they are there presented as definitely beliefs from the 'lower deck', from the seamen not the officers, the non-educated classes.(It would be a difficult business to extract all the mentions of luck from O'Brian's twenty-volume sequence of 'Aubrey and Maturin' novels, but I note an especially prominent statement in *The Ionian Mission* (1982), chapter 9, which distinguishes 'luck' carefully from 'chance, commonplace good fortune', and calls it 'a different concept altogether, one of an almost religious nature'.)

Tolkien probably thought that the very word 'luck' was Old English in origin (the *OED* insists that the 'ultimate etymology ... is obscure', but see the discussion on p. 145 below); and that once again ancient belief had survived into modern times unnoticed (just like hobbits). As with his riddles, Bilbo's 'luck' makes him seem more at home in the fairy-tale world, without being at all inconsistent with his modern English nature.

Bilbo's other quality, meanwhile, as noted by Thorin above, is courage, as he is to show again and again. But it is a significantly different type or style of courage from the heroic or aggressive style of his companions and their allies and enemies. Bilbo remains always unable to fight trolls, shoot dragons, or win battles. At the Battle of the Five Armies, even after he has grown in stature as far as is at all possible, Bilbo stays 'quite unimportant ... Actually I may say he put on his ring early in the business, and vanished from sight, if not from all danger'. However, after Gollum and his escape from the goblins, Bilbo does show that he has a kind of courage, and one which is comparable with and even superior to that of the dwarves. Now he has the ring, should he not 'go back into the horrible, horrible tunnels and look for his friends'? 'He had just made up his mind that it was his duty, that he must turn back' when he hears the dwarves arguing; they are arguing about whether they should turn back and look for him, and one of them at least says no: 'If we have got to go back now into those abominable tunnels to look for him, then drat him, I say'. Gandalf, of course, might have made them change their minds, but Bilbo is here for the first time shown as actually *superior* to his companions. His courage is not aggressive or hot-blooded. It is internalized, solitary, dutiful – and distinctively modern, for there is nothing like it in *Beowulf* or the Eddic poems or Norse saga. Just the same, it is courage of a sort, and even heroes and warriors ought to come to respect it.

The dwarves do indeed start to respect Bilbo from this point on, and Tolkien marks the stages through which this grows. In chapter 6, 'Bilbo's reputation went up a very great deal with the dwarves'. In chapter 8, 'Some of them even got up and bowed right to the ground before him', while in chapter 9 Thorin 'began to have a very high opinion [of Bilbo] indeed'. In chapter 11 he has more spirit left than the others, and by chapter 12, 'he had become the real leader in their adventure'. None of this stops the dwarves from returning to their earlier opinion of him – 'what is the use of sending a hobbit!' – and much of the time he reverts to being a passenger, as in the scenes with Beorn. But Bilbo's kind of courage is increasingly insisted on, always in scenes of solitude, always in the dark. Bilbo kills the giant spider 'all alone by himself in the dark', and it makes him feel 'a different person, and much fiercer and bolder'. After he has done it he gives his sword a name, 'I shall call you *Sting*', something much more likely for a saga-hero to do than for a modern bourgeois. His great moment, however, is to go on by himself in the dark tunnel after he has heard the sound of Smaug the dragon snoring:

> Going on from there was the bravest thing he ever did. The tremendous things that happened afterward were as nothing compared to it. He fought the real battle in that tunnel alone, before he ever saw the vast danger that lay in wait.

In all these ways Tolkien insists that Bilbo, or 'Mr. Baggins' as he is still often called, remains a person from the modern world; but that people from that world need not feel entirely alien in or inferior to the fairy-tale world.

Philological fictions

One aspect of the structure of *The Hobbit*, then, consists of Bilbo becoming more and more at home in the world of fairy-tale, in Middle-earth as it was to become. Another aspect, though, and one which Tolkien was uniquely qualified to create, consists simply of making that world more and more familiar: one might say, of making it up, though Tolkien himself might have rejected that description. Much of *The Hobbit* works, as has been said, by simply introducing a new creature (Gollum, Beorn, Smaug), a new species (dwarves, goblins, wargs, eagles, elves), or a new locale (the Misty Mountains, Mirkwood, Laketown), generally one or two to a chapter. Some of these innovations are inventions. There is no known source for Gollum other than Tolkien's own mind; it was his idea, and a brilliant one, to mark Gollum out by his strange use of pronouns. After his very first remark, 'I guess it's a choice feast', Gollum never again, in *The Hobbit* – *The Lord of the Rings* is a different matter – uses the word 'I'. He always calls himself 'we' or 'my preciouss'. He never says 'you' either, though, as with the trolls' distinctive speech, printers have done their best to 'make sense' of the abnormal, for instance by quietly rewriting 'we' as 'ye'. (Proof-reading errors and printers' errors continued to vex *The Hobbit* for many years. Only in very recent editions has the contradiction over 'Durin's Day' been resolved, which owners of earlier editions will find at the end of chapter 3 ['last moon of Autumn'] and the start of chapter 4 ['first moon of Autumn']. It should be 'last'.) Gollum's consistent verbal oddity gives a distinctive sense of personality, or lack of personality, which is entirely original. Similarly, though Tolkien said, or is said to have said, many years later that the giant spiders were a borrowing from Germanic legend, this is not true. They too are purely Tolkienian.

Gollum and the spiders are the exception, however. Most of Tolkien's creations in *The Hobbit* as in *The Lord of the Rings* are the product of Tolkien's professional discipline. The 'wargs' are a very plain case. There is a word in Old Norse, *vargr*, which means both 'wolf' and 'outlaw'. In Old English there is a word *wearh*, which means 'outcast' or 'outlaw' (but not 'wolf'), and a verb *awyrgan*, which means 'to condemn', but also 'to strangle' (the death of a condemned outcast), and perhaps 'to worry, to bite to death'. Why did Old Norse feel the need for another word for 'wolf', when they had the common word too, *úlfr*? And why should Old English give the word somehow a more eerie and less evidently physical sense? Tolkien's word 'Warg' clearly splits the difference between Old Norse and Old English pronunciations, and his concept of them – wolves, but not just wolves, intelligent and malevolent wolves – combines the two ancient opinions.

Beorn is another case in point. Here one might imagine Tolkien working a slightly different way. He had to teach the Old English poem *Beowulf* probably every year of his working life, and one of the elementary data about that poem (like most things about the poem, it took half a century to be noticed) is that the hero's name means 'bear': he is the bee-wolf, the ravager of the bees, the creature who steals their honey, hence (as every reader of *Winnie the Pooh* would recognize), the bear. Beowulf however, though he is immensely strong and a keen swimmer, both ursine traits (for polar bears in particular are semi-amphibious), remains human all the way through his story, with only very occasional hints that there may be something strange about him. His adventures are paralleled, though, in an Old Norse work for other reasons to be connected with *Beowulf*, *The Saga of Hrolf Kraki*, sometimes called *The Saga of King Hrolf and his Champions*. The head of King Hrolf's champions is one Böthvarr Bjarki, a clear analogue to Beowulf in what he does. Böthvarr is an ordinary name (it survives in the Yorkshire village of Battersby), but his nickname Bjarki means 'little bear'. Since his father's name is Bjarni (which means 'bear') and his mother's is Bera (which means 'she-bear'), it is pretty clear that Böthvarr is in some way or other a bear: in fact, a were-bear. Like many Old Norse heroes he is *eigi einhamr*, 'not one-skinned'. In the climactic battle he turns into a bear, or rather projects his bear-fetch or bear-shape out into the battle – till he is foolishly disturbed and the battle lost.

Tolkien put these pieces together – all of them, note, completely familiar to any *Beowulf* scholar, let alone one of Tolkien's eminence. If there is one thing clear about Beorn in *The Hobbit*, it is that he is a were-bear: immensely strong, a honey-eater, man by day but bear by night, capable of appearing in battle 'in bear's shape'. His name, Beorn, is the Old English 'cognate', or equivalent, of Böthvarr's father's, Bjarni, and in Old English it means 'man': but it used to mean 'bear', taken over and humanized just like, for instance, ordinary modern English 'Graham' (< 'grey-hame' < Old English *græg-háma = 'grey-coat' = 'wolf'). Yet as with the 'Tally of the Dwarves' Tolkien went beyond these merely verbal puzzles to ask himself, given all the data above, what would a were-bear actually be *like*? And the answer is Beorn, that strange combination of gruffness and good-humour, ferocity and kind-heartedness, with overlaying it all a quality which one might call being insufficiently socialized – all caused, of course, by the fact that he has 'more than one skin', is 'a skin-changer'. Gandalf insists on this duality from the beginning: 'He can be appalling when he is angry, though he is kind enough if humoured'; and it is kept up throughout, till they find the goblin-head and warg-skin nailed up outside his house: 'Beorn was a fierce enemy. But now he was their friend'. He remains a conditional sort of a friend, of course, as the dwarves would have found out if they had dared to take his ponies into Mirkwood. Beorn comes from the heart of the ancient world that existed before fairy-tale, a merciless world without a Geneva Convention. The surprising and charming thing about him, perhaps, and by no means inconsistent with his origins, is that he is at the same time a vegetarian, a model of co-operative ecology, and readily amused. In Beorn *Beowulf* and *Hrolfs saga* have been assimilated and naturalized.

Tolkien took not only riddles, and characters, but also settings from ancient literature. In another poem from the *Elder Edda*, the *Skírnismál*, there is a stanza which seems to have been as suggestive for him as the stanzas from the *Dvergatal* mentioned above. Just before it the god Freyr, passionately in love with a giantess, has decided to send his servant Skírnir to woo her for him, lending him his horse and his magic sword to help him. With heroic resignation Skírnir says – to the horse, not to Freyr:

'Myrct er úti, mál qveð ec ocr fara
úrig fiöll yfir,
þyr[s]a þióð yfir;
báðir við komomc, eða ocr báða tecr
sá inn ámátki iötunn.'

I translate, keeping as close to the original as possible:

'It is mirk outside, I call it our business to fare
over the rainy mountains,
over the tribes of thyrses;
we will both come back, or he will take us both,
he the mighty giant.'

It was characteristic of Tolkien in a way to ignore contexts, to seek suggestion instead in words, or names. Here he makes no use of Freyr, or Skírnir, or love for giant maidens, but he seems to have asked himself, 'what does *úrig* really mean? And what are these "tribes of thyrses"'? One answer to the last question is that they are a kind of orc – there is an Old English compound word *orc-þyrs*, which suggests that orcs are the same as thyrses. As for *úrig*, the German editors of the poem suggest as translations 'damp, shining with wet'. Tolkien seems to have preferred 'misty', with its suggestion of hidden landscapes. In *The Hobbit* Bilbo does exactly what Skírnir says he is going to do: he crosses the Misty Mountains, and passes over the tribes of orcs. But both are brought into sharp focus, instead of being forever on the edge of meaning, as in the Norse poem.

Tolkien derived Mirkwood in exactly the same way. *Myrcviðr* is mentioned several times in the Eddic poems. The Burgundian heroes ride through it, *Myrcvið inn ókunna*, 'Mirkwood the unknown',on their disastrous visit to Attila the Hun. Hlöthr the Hun claims it as part of his patrimony from his Gothic half-brother in the poem *The Battle of the Goths and Huns, Hrís þat it mæta, er Myrcviðr heitir,* 'the splendid forest that is called Mirkwood' – the poem forms part of *The Saga of King Heidrek* already mentioned. There seems to be general agreement among Norse writers that Mirkwood is in the east, and forms a kind of boundary, perhaps between the mountains and the steppe. But once again it is never brought into focus. Tolkien reacted, again, by bringing it into focus; by making it 'unknown', and almost literally pathless; by keeping it as a place one has to go through to get to a destination in the east; but also by populating it with elves.

He had, as we now know, been creating an elvish world and an elvish mythology for more than twenty years before *The Hobbit,* in the string of tales which were to become *The Silmarillion,* and which have been published in much greater detail in successive volumes of *The History of Middle-earth.* In 1937, though, he used these sparingly, mentioning them only with reference to Elrond in chapter 3, 'one of those people whose fathers came into the strange stories before the beginning of History', to 'the language that [Men] learned of elves in the days when all the world was wonderful' in chapter 12, and most of all in the long paragraph discussing the Wood-elves, High Elves, Light-elves, Deep-elves and Sea-elves in chapter 8. Tolkien drew his immediate inspiration for the Wood-elves of *The Hobbit* from, once again, a single passage from the Middle English romance *Sir Orfeo,* his complete translation of which was to appear many years later, in 1975. This contains a famous section in which King Orfeo, wandering alone and crazy in the wilderness after his wife has been abducted by the King of Faerie – the romance is a thoroughly altered version of the Classical myth of Orpheus and Eurydice – sees the fairies riding by to hunt. Tolkien's version of the lines goes:

> There often by him would he see,
> when noon was hot on leaf and tree,
> the king of Faerie with his rout
> came hunting in the woods about
> with blowing far and crying dim,
> and barking hounds that were with him;
> yet never a beast they took nor slew,
> and where they went he never knew.

The first sign of the elves in chapter 8 of *The Hobbit* is the flying deer which charges into the dwarves as they try to cross the water of oblivion in Mirkwood. After it has leapt the stream and fallen from Thorin's arrow:

> they became aware of the *dim blowing of horns* in the wood and the sound of dogs baying far off. Then they all fell silent; and as they sat it seemed they could hear the noise of a great hunt going by to the north of the path, though they saw no sign of it.

Orfeo's hunt is 'dim' because it is not clear he is in the same world as the fairies, who chase beasts but never catch them. The dwarves' hunt is 'dim', more practically, because they are after all in *Mirk*-wood and cannot see or even hear clearly. But the idea is the same in both places, of a mighty king pursuing his kingly activities in a world forever out of reach of strangers and trespassers in his domain. Tolkien expanded this very much, with ideas both from his own mythology (the underground fortress) and from traditional fairy-tale (the fairies who disappear whenever strangers try to intrude on them), but he continued to use the same technique as with riddles and Beorn and dwarf-names and place-names: he took fragments of ancient literature, expanded on their intensely suggestive hints of further meaning, and made them into coherent and consistent narrative (all the things which the old poems had failed, or never bothered, to do).

There is one final obvious use of old heroic poetry in *The Hobbit*, this time one which shows Tolkien especially clearly playing with anachronism, with the contrast of old and new: Bilbo's conversation with Smaug. For Tolkien's taste there were too few dragons in ancient literature, indeed by his count only three – the Miðgarðsorm or 'Worm of Middle-earth' which was to destroy the god Thor at Ragnarök, the Norse Doomsday; the dragon which Beowulf fights and kills at the cost of his own life; and Fafnir, who is killed by the Norse hero Sigurð. The first was too enormous and mythological to appear in a story on anything like a human scale, the second had some good touches but remained speechless and without marked character (though Tolkien did take from *Beowulf* the idea of the thief stealing a cup, and then returning, eventually in a company of thirteen). For the most part, though, Tolkien was left with the third dragon, Fafnir. In the Eddic poem *Fáfnismál* Sigurð stabs it from underneath, having dug a trench in the path down which it crawls – this is perhaps one of the 'stabs and jabs and undercuts' which the dwarves mention while they are discussing 'dragon-slayings historical, dubious, and mythical' in chapter 12 – but Fafnir does not die at once. Instead, for some twenty-two stanzas the hero and the dragon engage in a conversation, from which Tolkien took several hints.

The first is that in the Eddic poem Sigurð, to begin with, will not give his name, but replies riddlingly, calling himself both motherless and fatherless. Tolkien entirely remotivates this, explaining 'This of course is the way to talk to dragons ... No dragon can resist the fascination of riddling talk'. Sigurð's motive was that Fafnir was dying, and 'it was the belief in old times that the word of a dying man had great power, if he cursed his enemy by name'. But then the Eddic poem is, as often, a disappointment to a logical mind, for Sigurð does give his name very shortly after this, and Fafnir indeed seems to know all about him. Tolkien used the start of the conversation, then, and ignored its later development. He took a second hint from Fafnir's wily and successful attempt to sow discord between his killers, for Fafnir gives Sigurð unsought advice: 'I advise you, Sigurð, if you will take the advice, and ride home from here . . . Regin betrayed me, he will betray you, he will be the death of both of us'. In the same way Smaug tells Bilbo to beware of the dwarves, and Bilbo (with less reason than Sigurð) is for a moment taken in. There is a third hint after the dragon is dead, for Sigurð, tasting the dragon's blood, becomes able to understand bird-speech, and hears what the nut-hatches are saying: that Regin does indeed mean to betray him. In *The Hobbit*, of course, it is the thrush who proves able to understand human speech, not the other way round, and his intervention is fatal to the dragon, not to the dwarves. One can say only that Tolkien was well aware of the one famous human-dragon conversation in ancient literature, and admired the sense it creates of a cold, wily, superhuman intelligence, an 'overwhelming personality', to use Tolkien's entirely modern terminology. However, as often, Tolkien took the hints, but felt he could improve on them.

Much of the improvement comes from a kind of anachronism, which as so often in *The Hobbit* creates two entirely different verbal styles. Smaug does not, initially, talk like Beorn, or Thorin, or Thranduil the elf-king, or other characters from the heart of the heroic world. He talks like a twentieth-century Englishman, but one very definitely from the upper class, not the bourgeoisie at all. His main verbal characteristic is a kind of elaborate politeness, even circumlocution, of course totally insincere (as is often the case with upper-class English), but insidious and hard to counter. 'You seem familiar with my name', says Smaug, with a hint of asperity – being 'familiar' is low-class behaviour, like calling people by their first names on first meeting – 'but I don't seem to remember smelling you before'. Smaug could be a colonel in a railway carriage, spoken to by someone to whom he has not been properly introduced, and freezing him off with hauteur. He goes on with a characteristic mix of bluntness and the pretended deference which indicates offence: 'Who are you and where do you come from, *may I ask*?' (my emphasis). Bilbo then launches into his riddling introduction, but when Smaug talks at length again he has become in his turn familiar, even colloquial: 'Don't talk to me!' (this means, 'Don't try to fool me!'). 'You'll come to a bad end, if you go with such friends' ('friends' is entirely sarcastic). '*I don't mind* if you go back and tell them so from me' (my emphasis again: Smaug is still talking casually, but the understatement is clearly contemptuous). As he oozes confidentially on, his speech fills up with interjections, 'Ha! Ha! . . . Bless me! . . . eh?', and with further roundabout mock-courtesy, 'you may, perhaps, not altogether waste your time . . . I don't know if it has occurred to you that . . .' This is nothing like Fafnir, or Sigurð, or indeed any character from epic or saga, but it is convincingly dragonish: threatening, but cold, and horribly plausible. It is no wonder Bilbo is 'taken aback'.

However, this is not the only speech-mode Smaug has available. When Bilbo finally mentions to him the heroic motive of 'Revenge' – and Bilbo throughout the conversation talks in a much more elevated style than is usual for him – Smaug replies more archaically and more heroically than anyone has done in *The Hobbit* so far. 'I have eaten his people like a wolf among sheep, and where are his sons' sons that dare approach me . . . My armour is like tenfold shields, my teeth are swords, my claws spears, the shock of my tail a thunderbolt'. His language here approaches that of the Old Testament, and it is matched by the narrator's in describing him. After Bilbo's first theft, when Smaug wakes and finds he has been robbed, 'The dwarves heard the awful *rumour* of his flight' – 'rumour' here has the distinctly old-fashioned sense of 'far-off noise', not the weak modern one of 'gossip'. A couple of times Tolkien uses the device of substituting adjectives for adverbs, '*Slow and silent* he crept back to his lair ... floated *heavy and slow* in the dark like a monstrous crow', again creating an antique effect. Smaug's last boast to himself, at the end of chapter 12, 'They shall see me and remember who is the real King under the Mountain', uses the archaic third-person 'shall' of warriors' boasts in Old Norse and Old English, now condemned or marked as abnormal by modern school-grammarians. Smaug in fact seems to have a foot, or a claw, in two worlds at once. And in this at least he is like Bilbo the hobbit.

The clash of styles

Getting rid of Smaug remained, perhaps, Tolkien's major plotting problem in *The Hobbit*. His ancient sources were not much use to him. Odin's son Viðar kills the Fenriswolf by putting one foot on its lower jaw, seizing the upper jaw, and tearing it in half. There seems no likelihood of anyone in Middle-earth following suit. Sigurð's 'undercut' against Fafnir was too obvious to be used again, and Beowulf's self-sacrificing victory and death would involve creating a 'warrior', a character undeniably and full-time heroic, difficult to fit into the company of Bilbo. Tolkien solved his problem, as often, by a kind of anachronism, in the figure of Bard.

In some ways Bard is a figure from the ancient world of heroes. He prides himself on his descent, from Girion Lord of Dale. He re-establishes monarchy in Laketown, which till then seems to have been a kind of commercial republic. The proof of his descent lies in an inherited weapon which he speaks to as if it were sentient, and as if it too wanted vengeance on its old master's bane: 'Black arrow! ... I had you from my father and he from of old. If ever you came from the forges of the true king under the Mountain, go now and speed well!' And it is this arrow, of course – shot by Bard, but directed by the thrush, and ultimately by Bilbo – which kills the dragon, in a way not entirely dissimilar to Sigurth or Beowulf.

The death of Smaug is, however, presented for the most part and up to the final shot in a way which seems much more modernistic. It is above all a crowd scene. When the dragon-fire first appears in the sky, what Bard does is not prepare his own armoury, like Beowulf, but start to organize a collective defence, like a twentieth-century infantry officer. He has the whole town filling pots with water, readying arrows and darts, breaking down the bridge – the Middle-earth equivalent of digging trenches, collecting ammunition, organizing damage-control parties. Smaug is met by a fortified position and by volley-fire, with Bard running to and fro 'cheering on the archers and urging the Master to order them to fight to the last arrow'. The last word shows up the mixed nature of the scene, for the phrase one might expect is 'fight to the last *round*', a phrase from an era of musketry. In the same way 'there was still a company of archers who held their ground among the burning houses'. 'Hold one's ground' is another modernistic phrase, suggesting maps and front lines – the Old English version of it would be something like 'hold one's stead', i.e. the ground one stands on. The whole scene in fact, though transmuted into an era of bows and arrows, seems more like the First World War which Tolkien himself fought in than any legendary battle from the Dark Ages. Though victory does in the end turn on a single man and an ancestral weapon, the vigour of the description comes from collective action, from fore-thought and organization: in a word, from discipline.

I have commented elsewhere on the nineteenth-century idealiz-ation of this quality as the most prized of British imperial virtues (*Road*, 1992), and Tolkien was no stranger to it in real life. When, in his *Beowulf* lecture of 1936, he mentioned men in the present day 'who have heard of heroes and indeed seen them', he must have been referring to his own war service – I have no doubt that he knew, as a matter of regimental pride, that his own regiment, the Lancashire Fusiliers, won more Victoria Crosses (seventeen) during the First World War than any other. But when one talks of modern war-heroes, and then of ancient ones, the contrast of styles is very marked, the latter (for instance) having almost no conception of concern for others – no one is ever praised in saga literature, let alone decorated, for rescuing the wounded under fire – the former (by convention) without the personal and self-aggrandizing motives which so often, in epic or saga, come over nowadays as boastfully immodest. And yet, Tolkien must have thought, was there after all *no* connection, no connection at all? Could the relationship between Dark Age battles and the First World War not perhaps be like the one between Gollum and Bilbo: different on the surface, with a deeper current of similarity? That is what seems to be the case with Bard.

Superficial clash of styles leading to a deeper understanding of unity is in the end the major theme (even the major lesson) of *The Hobbit*. The superficial clash is exploited comically from the beginning, as when Bilbo's 'business manner' runs into the narrator's, and into Thorin's, in chapters 1 and 2. Bilbo speaks from the heart of the bourgeois world when he says obstinately, 'I should like to know about risks, out-of-pocket expenses, time required, and remuneration, and so forth', and the narrator immediately mocks him by putting the commercial language into plain speech – 'by which he meant: "What am I going to get out of it? And am I going to come back alive?"' Thorin then trumps even this with his letter which says, in a parody of business English, 'Terms: cash on delivery, up to and not exceeding one fourteenth of total profits (if any); funeral expenses to be defrayed by us or our representatives, if occasion arises and the matter is not otherwise arranged for'. Words and phrases like 'cash on delivery', 'profits', 'defrayed', were not and could not be used in medieval times (the word 'profit' is not even recorded in its modern sense by the *OED* till 1604). But on the other hand few modern contracts qualify profits with the gloomy phrase 'if any', or assume the likelihood that there will be *no* funeral expenses, because the contracting party or parties will have been eaten (though Beowulf says exactly that in lines 445–55 of the epic). Even the signature on the letter, 'Thorin & Co.', is ambiguous. Nothing could be more familiar in modern commerce than the '& Co.'. But Thorin's 'Co.' is not a *limited* company but a company in the oldest sense – fellow-travellers, messmates. In this initial clash Bilbo's deliberately grown-up style loses hands down. It seems pompous, evasive, self-deceiving, readily exposed by the dwarvish concentration on real probabilities.

After that, one might say, the styles see-saw. Thorin's is still up when they arrive in Laketown and he announces himself with genuine pomp as 'Thorin son of Thrain son of Thror King under the Mountain'. Fili and Kili get the same treatment, 'The sons of my father's daughter . . . Fili and Kíli of the race of Durin', but Bilbo is left as an anti-climax, 'and Mr. Baggins who has travelled with us out of the West'. One should note that Laketown is itself a place of clashing styles, in which there are at least three attitudes present: the Master's wary scepticism, similar to Bilbo's at the start, but extending in the case of the younger people to refusal to believe in any old tales about dragons at all; a counterbalancing and equally foolish romanticism, based on 'old songs' not very well understood, in which the dragon may exist but is no longer to be feared; and the grim and unpopular views of Bard, balanced between the two. Laketown, at the centre of the book, functions as another primarily hostile image of modernity, against which Thorin and the dwarves seem both splendid and realistic.

But then the stylistic see-saw tilts the other way. When Thorin launches into another magniloquent speech at the start of chapter 12, which contains within it the epic 'now is the time' formula, the narrator cuts it down with 'You are familiar with Thorin's style on important occasions', and Bilbo cuts it down further with a mix of plain speech and sarcastic exaggeration: 'If you mean you think it is my job to go into the secret passage first, O Thorin Thrain's son Oakenshield, may your beard grow ever longer . . . say so at once and have done!' Dwarvish rhetoric, and dwarvish splendour, are re-established by the sight of the treasure, which fills even Bilbo 'with enchantment and with the desire of dwarves'; but this too is kept in check by Bilbo's reactions. His *mithril* mail-coat and jewelled helmet ought to transform the hobbit even more than did the naming of Sting, but though he appreciates them, he cannot help putting himself back in a Hobbiton context: 'I feel magnificent . . . but I expect I look rather absurd. How they would laugh on the Hill at home! Still I wish there was a looking-glass handy!'

The final confrontation of styles comes, however, in chapters 15 and 16. Chapter 15, 'The gathering of the clouds', reaches a pitch of archaism higher than anything so far encountered. The raven Roäc son of Carc speaks with impressive dignity; Thorin's challenge repeats his titles, and is backed up by a newly aggressive version of the dwarves' song from chapter 1; the chapter leads on to the parley between Thorin and Bard, so archaically put, so full of rhetorical questions and grammatical inversions, that it remains quite hard to follow. One thing it does convey impressively is the difficulty of negotiation once issues of honour are involved. Much of the chapter would fit quite easily into situations from the Icelandic 'Sagas of the Kings'. But in the next chapter Bilbo takes a hand, and he does so with a return to the 'business manner' which was so unsuccessful at the start. Handing over the Arkenstone to Bard and the Elvenking, he says, 'in his best business manner': 'Really you know . . . things are impossible. I wish I was back in the West in my own home, where folk are more reasonable'. And with this he produces – from a pocket in his jacket, which he is still wearing over his mail – his original letter from 'Thorin & Co.'. His next proposal to them dwells on the exact meaning of 'profits', and uses words like 'claims' and 'deduct', all part of the vocabulary of the modern (Western) world and quite unknown to the ancient (Northern) one. But by this stage Bilbo has reverted all the way to his origins, and is furthermore demonstrating its ethical superiority. He rejects the suggestion of the Elvenking that he should stay with them in honour and safety, one should note, out of a purely private scruple, his word to Bombur, who would get the blame if he did not return. While this has some ancient Classical precedents – one thinks of Regulus returning to the Carthaginian torturers after having advised the Romans not to ransom him or his men – it is essentially kindly, un-aggressive, anti-heroic: though at the same time, like Bilbo deciding to go back into the goblins' tunnels, or down the tunnel to Smaug, undeniably brave. It is at this moment that Gandalf reappears, to ratify Bilbo's decision, re-establish him as 'Mr. Baggins', and send him off to dream not of treasure but of eggs and bacon.

Thorin then drops to his lowest point of the see-saw with his cursing of Bilbo, when Bilbo punctures dwarvish greeting formulas in much the same way that Gandalf punctured his own at the beginning: 'Is this all the service of you and your family that I was promised, Thorin?' It takes the Battle of the Five Armies and his own heroic death to re-establish Thorin, and in these events Bilbo plays almost no part at all, except to say deflatingly, 'I have always understood that defeat may be glorious. It seems very uncomfortable, not to say distressing'. (This may be a private joke. The King Edward's School Song, which Tolkien must have had to sing repeatedly in his youth, is aggressive even by Victorian standards, and contains the lines: 'Oftentimes defeat is splendid, / Victory may still be shame, / Luck is good, the prize is pleasant, / But the glory's in the game'.)

Thorin's two final speeches, however, show a balance of ancient epic dignity and a modern wider awareness: on the one hand, 'I go now to the halls of waiting, to sit beside my fathers', on the other, recognition of 'the kindly West' and 'a merrier world'. But a final and absolutely precise balance is reached only when Bilbo and the surviving dwarves part, with completely antithetical speeches:

> 'If ever you visit us again [said Balin], when our halls are made fair once more, then the feast shall indeed be splendid!'
>
> 'If ever you are passing my way,' said Bilbo, 'don't wait to knock! Tea is at four; but any of you are welcome at any time!'

Visit / pass my way, splendid / welcome, feast / tea: the contrasts of words and behaviour are obvious and deliberate. Yet it is also perfectly obvious that beneath these contrasts, both speakers are *saying exactly the same thing*. As with Gollum and Bilbo, Bard the bowman and Bard the officer, the heroes of antiquity and the Lancashire Fusiliers, there is a continuity between ancient and modern which is at least as strong as the difference.

Bridging the gap

The thought above may take us back to rabbits, and to hobbits. Tolkien's hobbits are like rabbits in a way which few people suspect, but which he himself was almost uniquely qualified to observe, that is, in their etymological history (real or imagined). The word 'rabbit' is a strange one. Almost all of the names for the wild mammals of England have remained more or less the same for more than a thousand years. Words like fox, weasel, otter, mouse, hare, were virtually the same in Old English, respectively *fuhs, wesel, otor, mús, hasa*. 'Badger' is a relatively new word, from French, but the old word, *brocc*, is still used: in later life Tolkien was short with translators who did not realize that the Shire place-name Brockhouses meant a badger sett. Such words tend to be the same in other Germanic languages too, so that the German for 'hare' is *Hase*, the Danish *hare*, and so on. The reason, obviously, is that these are old words for creatures which have long been familiar. But 'rabbit' is not like that. The words for the animal in neighbouring languages are different, so German *Kaninchen*, French *lapin*, and so on. There is no Old English word for 'rabbit'. Again, the obvious reason is that rabbits are a relatively recent import into England, like mink, brought in first by the Normans as fur-bearing animals, eventually released into the wild. However, not one English person in ten thousand realizes that, nor do they care. Rabbits have been naturalized, have made their way into folk-tale and popular belief and children's story, from Alison Uttley's Little Grey Rabbit to Beatrix Potter's Benjamin Bunny. Now it seems as if they have always been there.

This is the fate which I think Tolkien would like for hobbits. His dwarves and elves are similar, in the age of their names and their wide distribution, to hares and foxes. Hobbits are (if one discounts the slender evidence of *The Denham Tracts*) imports, like rabbits. But perhaps in the end, or even, by art, in the beginning, they can be made to seem harmonious, to settle in, to look as if they had been there all along – the niche which Tolkien eventually claimed for hobbits, 'an *unobtrusive* but *very ancient* people' (my emphasis). Tolkien even found an etymology for hobbits, as the *OED* has failed to do for rabbits. His first words about them were, as has been said, 'In a hole in the ground there lived a hobbit'. Many years and many hundreds of pages later, in almost the last words of the last Appendix of *The Lord of the Rings*, Tolkien suggested that 'hobbit' might be a modern worn-down form of an unrecorded but perfectly plausible Old English word, *holbytla*. *Hol* of course means hole. A 'bottle' even now in some English place-names means a dwelling, and Old English *bytlian* means to dwell, to live in. *Holbytla*, then, = 'hole-dweller, hole-liver'. 'In a hole in the ground there lived a hole-liver.' What could be more obvious than that? It is not impossible that Tolkien, one of the great philologists, who knew more about Old English in the 1930s than almost anyone alive, might have had this etymology in his head, perhaps subconsciously, when he wrote the seminal sentence on the School Certificate paper, but I think it is unlikely. What is more likely is that Tolkien, faced with a verbal puzzle, did not rest till he had worked out a totally convincing argument for it; while even in creating words he did so with a very strong sense of what fitted English patterns and what did not.

These comments on the word 'hobbit' furthermore fit the concept of hobbits. They are above all anachronisms, novelties in an imagined ancient world, the world of fairy-tale and nursery-rhyme and what once lay behind them. They retain that anachronistic quality stubbornly to the end, smoking tobacco (an import from America unknown to the ancient North), and eating potatoes (another import from America, on which old Gaffer Gamgee is an authority). The scene in the *Two Towers* chapter, 'Of Herbs and Stewed Rabbit', in which the hobbit Sam cooks rabbit, wishes for potatoes, and promises in better days to cook Gollum that English favourite, 'fish and chips', is a cluster of anachronisms. And Tolkien was certainly aware of them, for in *The Lord of the Rings* he changed the alien word 'tobacco' into 'pipeweed', often referred to the equally alien 'potatoes' as the more native-sounding 'taters' or 'spuds' – and in the 1966 edition of *The Hobbit* cut the equally alien word 'tomatoes' out altogether, replacing it in Bilbo's larder with 'pickles' (see *Bibliography*, p. 30).

However, Tolkien kept the hobbits as anachronisms, because that was their essential function. The ways of creativity are difficult if not impossible to follow, and neat schemas are likely to be wrong in their neatness, if not their general direction. But one could say, with no doubt over-simple neatness, that Tolkien, like so many of the philologists of previous generations, was aware of the great gaps between ancient literature (like *Beowulf*) and its downgraded modern successors (like the tale of 'The Bear and the Water-carl'), as of the inadequacies of both groups in both quantity and quality; that he felt the urge to fill the gaps – not for nothing was his first unpublished attempt at an elvish mythology called 'The Book of Lost Tales'; that he wished also, when doing so, to give some hint of the charm and the fascination of the poems and stories to which he dedicated his professional life; and that he wanted finally to bridge the gap between the ancient world and the modern one. The hobbits are the bridge. The world they lead us into, Middle-earth, is the world of fairy-tale and of the ancient Northern imagination which lay behind fairy-tale, rendered accessible to the contemporary reader.

The qualities of Middle-earth, finally, are evident. Its inhabitants frequently present a challenge to modern values through their superior dignity, loyalty (Fili and Kili dying for Thorin, their lord and mother's brother), scrupulosity (Dáin honouring Thorin's agreement, though Thorin is dead), or all-round competence. On the other hand modern values, as represented by Bilbo, as frequently respond to the challenge by decisions taken internally, without witnesses, prompted by duty or conscience rather than concern for wealth or glory. Bilbo, and through him Tolkien's readers, can come to realize that they too have a birthright in Middle-earth, need not be totally cut off from it (even if orthodox literary history has tried to assert that they are).

Meanwhile, if there are two further qualities that may finally be asserted for Tolkien's version of Middle-earth, they are these: emotional depth, and richness of invention. The former is unusual, though not quite unparalleled, in a children's book. Few writers for children nowadays would dare to include the scene of Thorin's death, or have a quest end with such a partial victory: 'no longer any question of dividing the hoard', many dead including immortals 'that should have lived long ages yet merrily in the wood', the hero weeping 'until his eyes were red'. Nor would they venture on such themes as the 'dragon-sickness' which strikes both Thorin and the Master of Laketown, so that the one is 'bewildered' morally, by 'the bewilderment of the treasure', the other physically, fleeing with his people's gold to die of starvation 'in the Waste, deserted by his companions'. As for the unforgiving ferocity of Beorn, the unyielding both-sides-in-the-right confrontation of Thorin and the Elvenking, the grim punctilio of Bard, even Gandalf's habitual short temper, all these are far removed from standard presentations of virtue as thought suitable for child readers – no doubt one reason why the book has remained so popular.

Turning to richness of invention, perhaps all one need say here is that in *The Hobbit* Middle-earth retains a strong sense that there is far more to be said about it than has been. As Bilbo goes home, he has 'many hardships and adventures . . . the Wild was still the Wild, and there were many other things in it those days beside goblins': one would like to know what they were. When Smaug is killed the news spreads far across Mirkwood: 'Above the borders of the Forest there was whistling, crying and piping . . . Leaves rustled and startled ears were lifted.' We never learn whose ears they were, but the sense is there that Middle-earth has many lives and many stories besides the ones that have come momentarily into focus. The trick is an old one, and Tolkien learned it like so much else from his ancient sources, *Beowulf* and the poem of *Sir Gawain*, but it continues to work. It may have been a surprise to its publishers that a work as *sui generis* as *The Hobbit* should have been a popular success, but once it was a success there can have been no surprise in the clamour for a sequel. Tolkien had opened up a new imaginative continent, and the cry now was to see more of it.

CHAPTER II

❖═◉═❖

THE LORD OF THE RINGS (1):
MAPPING OUT A PLOT

Starting again

One of the most undeniable (and admirable), if least imitated qualities of Tolkien's eventual sequel, *The Lord of the Rings*, is the complex neatness of its overall design. It is divided into six 'Books' (the three *volumes* in which it usually appears were a publishing decision based on the cost of paper in post-war Britain). The first Book takes Bilbo's successor Frodo, with his three hobbit companions and eventually Strider, or Aragorn, to Rivendell. There he is joined by Gandalf and the rest of the 'fellowship' of the ring, that is, Boromir, Legolas, and Gimli. Their journey south, during which they lose Gandalf, takes up the second Book. At that point the company of eight splits up. Boromir is killed. Frodo and Sam set off to reach Orodruin and destroy the Ring. Pippin and Merry are captured by the orcs. Aragorn, Legolas, and Gimli pursue them. During the third, fourth, fifth and part of the sixth Book these three groups, still further supplemented (by the return of Gandalf) and subdivided (by the separation of Pippin and Merry), weave their paths in and out of each others' knowledge, the latter often partial or mistaken. (See the diagram on p. 104.)

Symmetry is, however, more than discoverable, it is unmistakable, if you look for it. Thus, it could be an accident that both Books I and II, in *The Fellowship of the Ring*, contain a second chapter which is largely explanation of the past building up to decisions about the future – and ending with much the same decision, that Frodo has to take the Ring to the Cracks of Doom. It probably *is* an accident that both Books I and II contain much the same number of scene-shifts and scenes of threat – some three of the latter (Old Forest, Barrow-downs and Weathertop against Caradhras, Moria and the orcs in Lórien), and four or five of the former, with Lórien juxtaposed against the house of Tom Bombadil as an asylum, a place of safety. But thereafter symmetry becomes increasingly detailed. Two groups of the Fellowship meet strangers in the wilderness, and are helped by them: Aragorn, Legolas and Gimli by Éomer the Rider, as they pursue the orcs and their hobbit-captives across the Rohan prairie; Frodo and Sam by Faramir the Ranger, as they struggle towards Mordor through the woods of Ithilien. The decision to free and assist the members of the Fellowship is in both cases disapproved by the helpful strangers' superiors, Théoden and Denethor. These two last are furthermore strongly parallel to each other: they are both old men who have lost their sons (Théodred, Boromir) and see Éomer and Faramir as doubtful replacements. They die at almost the same time, at or during the Battle of the Pelennor Fields. Each has a hall which is described in close detail, in IV/6 and V/1 respectively, and the two descriptions take on special point if compared with each other – as do the scenes of confrontation between Éomer and Aragorn and Faramir and Frodo in III/2 and IV/5. (Page-references are not always helpful in a work as often reprinted and repaginated as *The Lord of the Rings*. Where reference to the text may be valuable, I use accordingly Book [*not* volume] and chapter numbers. Here, for instance, IV/6 means chapter 6 of Book IV, 'The Window on the West', in *The Two Towers*, while

V/1 means chapter 1 of Book V, 'Minas Tirith', in *The Return of the King*.)

Meanwhile Merry and Pippin are clearly set antithetically to each other, with Merry joining the Riders and Pippin the defenders of Gondor, where each rises to much the same rank. All these points tend to set up a detailed cultural contrast between the Riders and the Gondorians, while at the same time there is a running cultural clash between Legolas the elf and Gimli the dwarf, as there is a clash of policies between Gandalf and Saruman (initially similar to and sometimes mistaken for each other). All the way through the later Books there is moreover a deliberate alternation between the sweeping and dramatic movements of the majority of the Fellowship, and the inching, small-scale progress of Frodo, Sam, and Gollum. The irony by which the latter in the end determines the fate of the former is obvious, remarked on by the characters and by the narrator. Tolkien furthermore went to great lengths to build in moments of connection, as when Legolas sees an eagle near the start of III/2, but does not find out that it was 'Gwaihir the Windlord' on an errand from Gandalf till three chapters later; or when Sauron is distracted from guarding against Frodo and Sam by the *palantír* in the hands of Aragorn. Tolkien also very carefully (and laboriously) created an exact day-by-day chronology for all *parties*, signalled in the text by such things as the changes of the moon. There is no doubt that Tolkien did all this, little doubt that he meant to, and no doubt again that the effects created of variety, contrast, and irony are in major part responsible for the book's phenomenal and never-equalled success.

Tolkien, however, had no idea of any of this when he began to write, nor indeed for quite unlikely stretches of time once he did get started. He may have felt himself in rather a quandary after the success of *The Hobbit*. *The Hobbit* itself had been published almost by accident, with a pupil who knew of its existence recommending it to a publisher's representative who encouraged him to send it to Stanley Unwin, and Unwin sr. then giving the typescript to his eleven-year-old son Rayner to read and report on (see *Bibliography*, pp. 7–8). Once it had come out, had been acclaimed, and Unwin had not unreasonably asked for a sequel, Tolkien must have wondered what to do. The texts he had on hand, and on which he had been working for twenty years already, were versions, in poetry and prose, of the complex of tales associated with the *Silmarillion* (a complex discussed in detail in chapter V below). He duly sent a selection of these in, from which Unwin made a further selection to pass on this time to an adult and professional reader, Edward Crankshaw. Crankshaw, however, faced with a collection of seemingly genuine ancient legends which made no concession whatsoever to novelistic convention, was baffled, and confessed as much. The story is told in detail by Christopher Tolkien in *The Lays of Beleriand*, pp. 364–7, but one thing that was clear from the start was that no 'Silmarillion' could possibly be seen as a sequel to *The Hobbit*. Told as much, Tolkien, we now know, began work on the sequel which was to turn into *The Lord of the Rings* some time between 16th and 19th December 1937, in the university's Christmas vacation.

Yet however neat the final product, at that point in late 1937, and for long afterwards, Tolkien had no clear plan at all, certainly nothing even remotely like the schema outlined at the start of this chapter. It is an interesting, and for any intending writer of fiction rather an encouraging experience, to read through the selections from Tolkien's many drafts now published in volumes VI-IX of *The History of Middle-earth* (*The Return of the Shadow*, *The Treason of Isengard*, *The War of the Ring*, and *Sauron Defeated*), and to note how long it was before the most obvious and seemingly inevitable decisions were made at all. Tolkien knew, for instance, that Bilbo's ring now had to be explained and would become important in the story, but he still had no idea of it as the Ring, the Ruling Ring, the Ring-with-a-capital-letter, so to speak: indeed he remarked at an early stage that it was 'Not very dangerous' (see *The Return of the Shadow*, p. 42). Another element arrived at early on was the character who would become Strider, the Ranger, but in several opening drafts this role of guard and guide is taken not by a man, still less by one of the Dúnedain, but by a weatherbeaten hobbit called Trotter, distinguished by his wooden shoes. Tolkien remained strongly attached to this character, and even more strongly attached to the name Trotter, though he was quite perplexed as to how to explain him. In *The Return of the Shadow* we see Tolkien wondering whether Trotter might perhaps be Bilbo in disguise; or maybe a relative, a cousin, one of the 'quiet lads and lasses' led off by Gandalf 'into the Blue for mad adventures'. Reading these drafts one often feels like saying, as Tolkien had done over the idea of fairy-hobbit marriages, 'This is, of course, absurd' (for all critics have 20/20 vision, in hindsight). However Christopher Tolkien notes that more than two years after his father started work on the sequel, he was still 'without any clear conception of what lay before him' (*The Treason of Isengard*, p. 18). 'Giant Treebeard' was at this stage hostile, and was the character responsible for the imprisonment of Gandalf, rather than Saruman, who had not yet appeared (*The Return of the Shadow*, p. 363). There was 'not a hint' of Lothlórien or of Rohan (*The Return of the Shadow*, p. 411), even by the time the Fellowship had reached Moria; Tolkien knew no more than his characters what lay the other side of the mountains. Perhaps the most surprising of the many surprises revealed by the early drafts is that in August 1939, with Tolkien about halfway through what would become Book II, of the eventual six Books, he thought that the work was about three-quarters done, see *The Return of the Shadow*, p. 370. It is as if he anticipated finishing not at the end of *The Lord of the Rings*, but at the end of *The Fellowship of the Ring*. The most determined hindsight, reading these drafts, can find no trace at all of the outline given above.

One critical factor in the development of the whole seems to have been the introduction of the Riders of Rohan, like Treebeard seen originally as enemies, allies of Sauron (*The Return of the Shadow*, p. 422) – Tolkien was indeed to keep this idea vestigially present in the completed work as a rumour, which Boromir indignantly rejects in II/2, but which is still present in the minds of Aragorn and his companions when they meet the Riders for the first time on the prairie at III/2. However, once they did appear, the Riders expanded the story markedly, and also gave Tolkien an easy way of tapping once more into the source-material of ancient literature. At about the same time he formed the idea of creating a set of linguistic correspondences within Middle-earth, and in the process providing a sensible explanation of the names used in *The Hobbit*. Tolkien knew (none better) that the dwarf-names he had used in *The Hobbit* came from Old Norse; but if one thought about it, it was clearly impossible that anything like these names could have survived from the far past of the Third Age. Old Norse is indeed an old language, but not so old that we cannot see its descent from something even older. As far back as Tolkien's Third Age whatever was the ancestor of Old Norse would be quite unrecognizable. The dwarf-names of *The Hobbit* must accordingly in strict logic be translations, and so must the hobbit-names; but in that case the real original hobbit-names and dwarf-names ought to have been related to each other in at least the same sort of way as modern English and Old Norse (which are in fact related, even quite closely related). The Riders could then be conceived of as being something linguistically in between hobbits and dwarves, as speaking (and in every detail except one, as being) Old English. Théoden realizes early on that there is some sort of connection between the hobbits and his people, a closer one than there is between the hobbits and the Northern men from whom the dwarves have borrowed their names and the language they use in public; his ancestors and theirs must at some time have lived in close association. Tolkien had worked out this set of relationships by about early 1942 (see *The Treason of Isengard*, p. 424), and could see his way at last to integrating it with the elvish languages and legends on which he had worked for so long already: this gave his story a clearer shape. However one thing which remains certain is that he was still not working from a plan, an overall design. He was writing his way into the story. Other great works have been written the same way, like Dickens's novels, composed and published in serial instalments – Tolkien's notes often look rather like Dickens's, with both writers in the habit of jotting down a string of possible names for a character till they struck one which seemed to fit. But Tolkien, even more than Dickens, had no conscious idea of where he was going. Seven months after starting work on *The Lord of the Rings*, he complained that he still had no story (*The Return of the Shadow*, p. 108). The amazing thing is that this did not stop him trying to write one.

Back-tracking

Tolkien did in fact have several resources when he began work in December 1937. One was the backlog of material which would in the end become the *Silmarillion*. As mentioned above, he had already sent some of this to Stanley Unwin, and though it had been rejected for separate publication, he could clearly continue to use it, as he had here and there in *The Hobbit*, to give a sense of depth and background to his main story. Thus Aragorn, in the chapter 'A Knife in the Dark' (I/11), not only sings a song of Beren and Lúthien, but also gives an extensive paraphrase of the legend concerning them, which had formed a major part of the package rejected by Unwin. Later Bilbo in Rivendell, in the chapter 'Many Meetings' (II/1), sings a song of Eärendil. Both poems were based on ones which Tolkien had already written and published separately, if only in university magazines of limited circulation. This indeed was a further resource available to Tolkien in 1937. Most of the more than a dozen poems in *The Hobbit* had been light-hearted or frivolous, like the elf-song in chapter 3 or the songs for taunting spiders in chapter 8, but some, especially the ongoing ballad which the dwarves start in chapter 1 and extend or modify according to their mood in chapters 7 and 15, had shown how poetry could be mixed in to narrative. Between 1923 and 1937 Tolkien had not only written but published a small body of poems which did not arise out of his *Silmarillion* legendarium, but which were available for re-use. However, his most important and unexpected resource in 1937, though it was not unconnected with the poems just mentioned, was a strong interest in place, and in place-names.

Place-names, like riddles and fairy-tales and nursery-rhymes, form yet another connection with antiquity in which Tolkien took strong personal interest. They were especially valuable to him for two reasons. One is that most people do not think much about names, but accept them as a given. They are accordingly unlikely to meddle with them, or change them except by the slow and natural processes of language change of which they are unconscious; which means that names may well contain unusually authentic testimony to history or to old tradition. Tolkien suggested to me once that the name of the village Hincksey, outside Oxford, might contain within it the name of the old hero Hengest, the founder of England (< *Hengestes-ieg*, 'Hengest's island'). He thought his own aunt Jane Neave's surname might be derived from the name of Hengest's dead leader, Hnæf. But another reason for taking an interest in names is that, unlike other words, they exist in a special relationship to what they refer to: obviously, one-to-one. It was said above, on p. 3, that 'The word is not the thing', but names are a lot closer to things than are other classes of word. If a name exists it offers a kind of a guarantee that what it labels must also exist. Names, especially names which are not strictly necessary, weight a narrative with the suggestion of reality. This may of course be just a device – a good example is the little elegy on Théoden King's horse in *The Return of the King*, which goes:

> Faithful servant yet master's bane,
> *Lightfoot's foal*, swift Snowmane.

Clearly we do not need to know the name of the father, or mother, of Snowmane, or indeed of Snowmane, a very incidental character. But giving *both* names, one completely extraneous to the story, is a sort of re-assurance. As said on p. 15 above, there are not many real names in *The Hobbit*, apart from those of the dwarves, and some of those listed there were added in later editions, but *The Lord of the Rings* is completely different. It is loaded down with names, personal names and place-names, the latter often transferred on to a map. They say a good deal about the way in which Tolkien began to work.

A sidelight on his methods and interests at this stage is cast by Tolkien's short story *Farmer Giles of Ham*. This was not published till 1949, but we know (see *Bibliography*) that it was heavily rewritten at about this time, and read to an Oxford college literary society in January 1938, a month after Tolkien began *The Lord of the Rings*. In it one can see Tolkien brooding, not only over nursery-rhyme, but also over the place-names of Oxfordshire and the neighbouring counties. The fictional 'Ham' of the title is the real village of Thame. Why should it be called Thame? Why is there an -h- in Thame, as in Thames, which no one ever pronounces? Should it not be Tame, and if so, what does Tame mean? Not far away from Thame is the equally real village of Worminghall, which seems on the face of it to mean 'the hall of the Wormings'. But what are Wormings? If 'worm' means dragon (as it often does in Old English), then might Wormings have something to do with a dragon – conceivably a *tame* dragon, since Tame is so close by? From thoughts like this Tolkien constructed his story of the wicked dragon Chrysophylax, who is bested by Farmer Giles with his sword Tailbiter, or Caudimordax, and who enables Farmer Giles to escape from the tyranny of the king of the Middle Kingdom, Augustus Bonifacius Ambrosius

Aurelianus Antoninus. The whole story is set in an imaginary past, the past of the 'Brutus books' referred to by the author of *Sir Gawain* (which Tolkien had co-edited thirteen years before), but its geography is perfectly realistic. The villages of Thame and Worminghall, and Oakley, which had its parson eaten, can all still be found close together on the map of Oxfordshire and Buckinghamshire – Thame is in Oxfordshire, the other two in Buckinghamshire; and the same map will show the small town of Brill in Buckinghamshire, once *Bree-hill – it was to be reshaped in *The Lord of the Rings* as Bree. The capital of the Middle Kingdom is not named, but is said to be 'twenty leagues' away, or sixty miles; it must be Tamworth, the ancient capital of Mercia, sixty-eight miles from Thame as the crow flies. The village of Farthingho, where we are told Farmer Giles's Little Kingdom maintained 'an outpost against the Middle Kingdom', is almost exactly on a line between them, one-third of the way measured from Thame. When Farmer Giles grumbles about the strange people who live far away, 'beyond the Standing Stones and all', he must mean the inhabitants of Warwickshire as opposed to his own Oxfordshire: the boundary between the two counties runs by the famous Rollright Stones. In *Farmer Giles* Tolkien was using place-names in a way which he had avoided in *The Hobbit*, but which he was to rely on in *The Lord of the Rings* (names like Brill, or Bree, or T(h)ame, or Farthingho, are on the face of it very different from The Hill or The Water, if not always so different historically). He was moreover taking a close interest in locality.

Tolkien used his new involvement with names in creating 'the Shire', with its elaborate map, relatively elaborate social structure, and elaborated history, all explained in the 'Prologue' to *The Lord of the Rings*. The Shire is indeed a brilliant invention, rubbing home the point that hobbits are just English people by its names, often strange-sounding (Nobottle, the Farthings) but usually real (there is a Nobottle in Northamptonshire, and a Farthingstone);

and by the very careful, point-for-point resemblance of its history to the traditional history of England, which extends even to both communities being founded by two brothers called 'Horse' – Hengest and Horsa for England, Marcho and Blanco for the Shire, but all four names are Old English words for the same animal. It also rationalizes some of Tolkien's anachronisms in *The Hobbit*, explaining that 'pipeweed' is the hobbits' only contribution to civilization, but no one knows where they got it from, and that a postal service is one of the few public functions exercised by the hobbits' minimal government, along with Shirriffs (sheriffs, or as Tolkien knew, 'shire-reeves'), the Mayor (another ancient office surviving into modern England, and as with sheriffs into America), and the Thain (Old English *thegn*, king's servant, now known to most people only from the Thane of Cawdor in *Macbeth*). But none of this solved Tolkien's underlying problem with story, with getting the story moving. What did was a poem he had published a few years before *The Hobbit*, which also springs from close engagement with locality, names and maps.

The poem was 'The Adventures of Tom Bombadil', which Tolkien published in *The Oxford Magazine* in 1934. Many years later it was to be reworked and to become the lead-title in the collection of poems published under that name as a sort of Christmas present for Tolkien's favourite aunt from 'Bag End', Jane Neave, but its 1934 version is somewhat different, provoking a narrative rather than (as in 1962) conforming to one already written. Both versions introduce Tom Bombadil without further explanation:

> Old Tom Bombadil was a merry fellow;
> bright blue his jacket was, and his boots were yellow.

Both also give Tom four adventures, or encounters with malignant powers. In the first he is pulled into the river by 'Goldberry, the Riverwoman's daughter'; in the second trapped by 'Willow-man'; in the third pulled down their hole by 'Badger-brock' and his family; in the last, as he gets home, he finds 'Barrow-wight' waiting for him behind the door:

'You've forgotten Barrow-wight dwelling in the old mound
up there a-top the hill with the ring of stones round.
He's got loose tonight: under the earth he'll take you!
Poor Tom Bombadil, pale and cold he'll make you!'

Tom reacts to all these adventures with complete confidence and simple imperatives, always obeyed: 'Go down! . . . let me out again . . . show me out at once . . . Go back to grassy mound . . . go back to buried gold and forgotten sorrow'. And then, the next morning, he goes back again, this time to capture Goldberry in his turn and take her off to be married. The beasts and bogies still cluster round his house in the night, tapping on the window-pane, sighing in the reeds, crying from the mound, but Tom ignores them all. Both versions end with him singing at sunrise:

sitting on the doorstep chopping sticks of willow,
while fair Goldberry combed her tresses yellow.

To readers of *The Oxford Magazine* in 1934 the poem must have seemed almost a nonsense-poem. What it does is to take the English landscape, perhaps the safest in the world, and to try to make it haunted. Tolkien did have a little to work on. Barrow-wights are familiar in Norse saga as ghosts, or more accurately walking corpses, coming out of their grave-mounds for vengeance on the living. There is little trace of this belief in English folklore, but against that, barrows are utterly familiar. Barely fifteen miles from Tolkien's study the Berkshire Downs rise from the Oxfordshire plain, thickly studded with Stone Age mounds, among them the famous Wayland's Smithy, from which a track leads to Nine Barrows Down. As he did elsewhere, Tolkien has taken one of the traces of old belief surviving in Norse, and anglicized it, transferring it to a locality he knew well. Meanwhile Goldberry is 'the Riverwoman's daughter', beautiful and charming herself, but connected with the hag who lurks like Grendel's mother 'in her deep weedy pool'. The folklore of hags has not been much studied, but Beowulfian scholars had at least heard of the malignant female river-deities whom some saw as a model for the *Beowulf*-poet. R.W. Chambers, a patron and supporter of Tolkien in his early years, had pointed to beliefs about Peg Powler, in the River Tees, and Jenny Greenteeth in the Ribble, as classic malignant water-hags.

What was missing in the 1934 version of the poem was the name of the river to which the Riverwoman belonged. In 1962 Tolkien was able to write it in as the Withywindle. This gives further clues as to how Tolkien was working, both with names and with locality. The description of the Withywindle itself, when the hobbits come upon it, is one of many brilliant passages of natural description in *The Lord of the Rings*. The hobbits find themselves coming down a 'deep dim-lit gully' which opens into a sunny valley:

> A golden afternoon of late sunshine lay warm and drowsy upon the hidden land between. In the midst of it there wound lazily a dark river of brown water, bordered with ancient willows, arched over with willows, blocked with fallen willows, and flecked with thousands of willow-leaves. The air was thick with them, fluttering yellow from the branches; for there was a warm and gentle breeze blowing softly in the valley, and the reeds were rustling, and the willow-boughs were creaking.

If Tolkien had left his study in Northmoor Road, walked back to the University Parks, crossed the Rainbow Bridge, and then walked along the other side of the river away from the town of

Oxford in the direction of the villages of Wood Eaton and Water Eaton – as no doubt he did – he would have seen virtually the same sight: the slow, muddy, lazy river fringed with willows. The real river, the one that flows into the Thames at Oxford, is the Cherwell. The *Oxford Dictionary of English Place-Names* gives a different derivation, but Tolkien was always capable of rejecting the advice of Oxford dictionaries. I think he derived the name from Old English **cier-welle*, the first element coming from *cier-ran*, 'to turn': so, 'the turning stream, the winding stream', which is what the Cherwell is (unlike the Evenlode not far away, the 'even-course', or the Skirfare in Yorkshire, the 'bright-runner', in which Tolkien's Leeds predecessor Professor Moorman had been drowned in 1919). Further down the Thames, furthermore, is Windsor, which may take its name from **windels-ora*, 'the place on the winding stream'. Finally, 'withy' is simply the old word for 'willow', frequent in English place-names, like the Warwickshire Withybrook. The Withywindle is a combination of the Cherwell itself, and words for its two main features, its willows and its slowly twisting course. We have no name for its resident hag, but it would only be sensible to see her as more passive, and perhaps more likely to have a human-friendly daughter, than child-eating Peg Powler, the spirit of the rapid River Tees, running down from the highest waterfall in England, High Foss. Willowman, meanwhile, though a 'grey thirsty spirit', 'filled with a hatred of things that go free upon the earth', works mostly by the power of sleep – he sends an overpowering drowsiness so he can trap or drown the hobbits. Such a power is notoriously at its height in long quiet summer afternoons, especially (some say) in river-bottom country like Oxford.

Finally Tom Bombadil himself was from his first conception a *genius loci*, a 'spirit of the place', the place being, as Tolkien remarked to Unwin (see *Letters*, p. 26), 'the (vanishing) Oxfordshire and Berkshire countryside'. The elves in *The Lord of the Rings* call him 'oldest and fatherless'; he is the one creature over whom the Ring has no power at all, not even to make invisible; but he could not defy Sauron permanently, for his power 'is in the earth itself', and Sauron 'can torture and destroy the very hills'. He is a kind of exhalation of the earth, a nature-spirit and once again a highly English one: cheerful, noisy, unpretentious to the point of shabbiness, extremely direct, apparently rather simple, not as simple as he looks. The fact that everything he says is in a sort of verse, whether printed as verse or not, and that the hobbits too find themselves 'singing merrily, as if it was easier and more natural than talking', make him seem, not an artist, but someone from an age before art and nature were distinguished, when magic needed no wizard's staff but came from words alone. Tolkien may have got the idea from the singing wizards of the Finnish epic the *Kalevala*, which he so much admired, and which he perhaps wished might also have an English counterpart.

The point one may take from these comments is that for the first nine chapters at least the action of *The Lord of the Rings does not move very far*. The hobbits get out of the Shire, true. But in the Old Forest, along the Withywindle, and on the Barrow-downs, they are still moving in a very familiar landscape, all of it within a day's walk of Tolkien's own study. Bree itself is modelled on the Buckinghamshire town of Brill. One thing Tolkien certainly knew about the latter is that its name is odd and philologically interesting, being composed of two elements, 'bree' and 'hill'. But 'bree' is only the Welsh word for 'hill'. The name suggests that the incoming English heard the word, used as a description, but thought it was a name and added their own description to it, creating 'Bree-hill': not very far removed, in origin, from 'The Hill' in *The Hobbit*, but of course felt quite differently now that it has become a name, and so in one-to-one relationship with what it represents as ordinary nouns are not. The name Bree bears witness also to Tolkien's other theory about place-names, mentioned in the Foreword: that people could still detect 'linguistic style' in them. Tolkien backed his hunch by giving the villages round Bree names of the same kind – Archet, from Welsh *ar chet*, 'the wood', Combe, from Welsh *cŵm*, 'valley'. He wanted the Bree area to feel slightly different from the Shire, and trusted to his readers' intuitions for the result. In the process, of course, he created a further sense of the variety and verisimilitude of Middle-earth. Much of his seemingly redundant activity in finding names and drawing maps paid off in this way.

Just the same, these factors together – Tolkien's expansion of the earlier Bombadil poem, his consequent confinement within a familiar imaginative space, the concern for locality signalled by the new interest in names inside and outside the Shire – explain why, for all Tolkien's efforts, *The Lord of the Rings* takes such a time to develop its major theme. The hobbits in particular have to be dug out, or winkled out, of no fewer than five Homely Houses before the voyage of the Ring starts in earnest. First there is Bag End; then the (really rather unnecessary) halt at Fredegar Bolger's house at Crickhollow; then the house of Tom Bombadil, then the *Prancing Pony*, and finally the house of Elrond. Furthermore much of the activity of the hobbits in these sections comes not from their adventures but from their recuperations: feasting with the elves in the Shire, hot baths in Crickhollow, singing with Tom Bombadil, singing again in the common room of the *Prancing Pony*, working their way through 'yellow cream, honeycomb, and white bread and butter', 'hot soup, cold meats, a blackberry tart, new loaves, slabs of butter, and half a ripe cheese', not to mention the elves' 'fruits sweet as wildberries and richer than the tended fruits of gardens' and Farmer Maggot's 'mighty dish of mushrooms and bacon'. It was a criticism of Tolkien's early drafts of Book I made by both Rayner Unwin and C.S. Lewis that Tolkien found it too easy, and too amusing, just to let the hobbits chatter on. Tolkien did his best to amend this (one can see him responding to the criticisms in a letter to Stanley Unwin in 1938), and the Old Forest and Barrow-wight sections have their own charm and their own power; but there is still a faint sense, after much reworking, that Tolkien was initially groping for a story, and keeping himself going with a sort of travelogue.

The clearest sign of this comes, perhaps, from the early use made of the Black Riders, the Nazgûl. The concept of the 'Ring-wraith', and indeed of the 'wraith' more generally, is an original and compelling one, with surprising modern resonance (as will be argued in the next chapter, on Tolkien's presentation of evil). But though the Riders appear frequently in the 'travelogue' section, from Bag End to Rivendell, they show relatively little of the force and meaning which they acquire later on. In their first appearance (in I/3) a Rider is seen 'sniffing' for Frodo, and Frodo feels an urge to put on the Ring. But nothing happens – it is an odd thing that when this description was first written the mysterious rider turned out in fact to be Gandalf (*The Return of the Shadow*, p. 47)! After that there is a second 'snuffling' scene, from which the hobbits are saved by the elves. The third time a Rider appears (I/4), he is deterred by no more than the difficulty of taking his horse down a steep bank. The 'long-drawn wail' which follows, 'the cry of some evil and lonely creature', is a feature which will be developed, but this time it has none of the effects of despair and demoralization which occur in later Books. On three occasions, we are told, the Riders try to get information from one person or another, from Gaffer Gamgee, Farmer Maggot, Barliman Butterbur, but though they are ominous, with their hissings and their ability to make hair stand on end, they seem to have no special powers. The two armed attacks they make, on the house at Crickhollow and on the *Prancing Pony*, do not amount to much – 'good bolsters ruined and all', says Butterbur. Nob drives off the Riders bending over Merry with a shout. There are only two scenes in the early sections where the Riders develop any of the supernatural power later ascribed to them, and one of those is again told rather than shown – Gandalf's description of the splinter from the Morgul-knife working closer to Frodo's heart, which would have turned him into 'a wraith under the dominion of the Dark Lord'. The attack on Weathertop at the end of I/11 does give a glimpse of what the Riders are in the other world, with the sight given by the Ring: white faces, grey hairs, haggard hands, something not skeletal but undying, the bitter and dangerous obverse of the long life enjoyed by Bilbo and endured by Gollum. But this is only a glimpse, and may have been written in at a late stage. Speaking purely tactically, one has to say that the Riders could have saved themselves a great deal of trouble later on by pressing their attacks home at this early point. The reason for their lack of development remains, once again, that to begin with Tolkien had no story. As he had done in *The Hobbit*, he was writing himself into the experience of Middle-earth, and using such material as he had already available.

One does not need to conclude that this was a failure, or a mistake. In later years Tolkien was to toy with the idea that these early chapters were in fact integrated to the main plot, that (for instance) Willow-man, the Barrow-wight, and the elementals who send the storm on Caradhras, were all operating under the command of the chief Ringwraith (*Unfinished Tales*, p. 348). This is not the impression they make at the time. Aragorn says of Caradhras that 'There are many evil and unfriendly things in the world that have little love for those that go on two legs, and yet are not in league with Sauron, but have purposes of their own', and Gimli agrees with him: 'Caradhras was called the Cruel, and had an ill name . . . when rumour of Sauron had not been heard in these lands'. Aragorn says earlier on, of Butterbur, that he lives 'within a day's march of foes that would freeze his heart, or lay his little town in ruin, if he were not guarded ceaselessly', but he never says what those foes are – trolls? ettins from the ettinmoors? orc-tribes? killer-Huorns? The scene with the wight is especially mysterious in that we never learn what it is that the wight intends to do, why it has dressed its captives in the gold of the buried dead, or why it has seemed in a way to reanimate its old victims (or itself?) in the bodies of the hobbits – for when Merry wakes up he thinks for a moment that he is a warrior killed long ago in battle against the Witch-king, who will eventually become the chief of the Nazgûl. Nevertheless the wight is not a wraith, the creatures of Caradhras seem perhaps to be mad Bombadils, the *genii* of a cruel and inhuman landscape, Butterbur's foes remain unglimpsed. And as with the mention of 'other things' in the Wild besides goblins, in *The Hobbit*, that is entirely to the good. When Tolkien drew his maps and covered them with names, he felt no need to bring all the names into the story. They do their work by suggesting that there is a world outside the story, that the story is only a selection; and the same goes for the hints of other creatures unaffected by and uninterested in the main plot. Middle-earth is different from its many imitators in its density, its redundancy, and consequently its depth, and Book I of *The Lord of the Rings* does a great deal to create that depth. What was needed next, though, was a greater degree of narrative urgency.

The Council of Elrond: character revealed

Urgency is supplied in large part by chapter 2 of Book II, 'The Council of Elrond'. The chapter is a largely unappreciated *tour de force*, whose success may be gauged by the fact that few pause to recognize its complexity. It breaks, furthermore, most of the rules which might be given to an apprentice writer. For one thing, though it is fifteen thousand words long, in it nothing happens: it consists entirely of people talking. For another, it has an unusual number of speakers present (twelve), the majority of them (seven) unknown to the reader and appearing for the first time. Just to make things more difficult, the longest speech, by Gandalf, which takes up close on half the total, contains direct quotation from seven more speakers, or writers, all of them apart from Butterbur and Gaffer Gamgee new to the story, and some of them (Saruman,

Denethor) to be extremely important to it later on. Other speakers, like Glóin, give direct quotation from yet more speakers, Dáin and Sauron's messenger. Like so many committee meetings, this chapter could very easily have disintegrated, lost its way, or simply become too boring to follow. The fact that it does not is brought about by two things, Tolkien's extremely firm grasp of the history (as earlier of the geography) of Middle-earth; and his unusual ability to suggest cultural variation by differences in mode of speech.

With more than twenty voices to deal with, demonstrating this latter point fully would be a long business, and I pick out accordingly only some of the most marked variations. Elrond is an immortal, and by far the oldest speaker present – Frodo is taken aback to realize he has been an eye-witness of events now legendary. It is only suitable, then, that his speech is strongly (though never incomprehensibly) marked by archaism, in particular by unusual use of word-order. Modern English, it may be said – Tolkien, as a Professor of English language, could at any time have expanded this brief account into a full lecture – has grown increasingly inflexible in its rules about word-order. It is rare for subjects not to precede verbs, or for objects and complements not to follow them; the old rule that the verb comes second, and so must change places with the subject if something else has taken first place, has almost vanished. But not quite, as one can see from sentences like 'Down came the rain' or 'Up went the umbrellas', both archaic, in a way, but at the same time completely colloquial. Elrond's speech is a kind of treasury of sub-rules of this nature. See for instance:

'This I will have as weregild for my father, and my brother'
(Elrond is quoting Isildur, from the far past: he uses the
archaic word 'weregild', and puts the grammatical object
in first place with '*This* I will have ...')

'Only to the North did these tidings come' (Elrond is speaking in his own person: he uses the archaic word 'tidings', which is still familiar however from the Christmas formula 'tidings of great joy'; he puts the adverbial phrase first, '*Only to the North . . .*', which he follows by inverting subject and verb, as in 'Down came the rain' constructions)

'From the ruin of the Gladden Fields . . . three men only came ever back' (Elrond is speaking in his own person again, but his word-order is odd, cp. more normal 'only three men ever came back')

'Fruitless did I call the victory of the Last Alliance?' (Elrond once more, but this time he has placed a grammatical complement first, 'Fruitless', and once again inverted subject and verb).

Elrond's archaism is consistent, achieved not just by vocabulary (the first resort of the amateur medievalist), but also by grammar. Though marked, it is never so obtrusive as to obscure meaning or make the speaker appear quaint. It serves to distinguish his speech from that of the others; to act as a continual reminder of his age; and to make a link with the similarly archaic speech of Isildur, when Gandalf also comes to quote this later on. Many critics have complained of Tolkien's archaic style in one section or another; they have failed to realize that he understood archaism far more technically than they ever could, and could switch it on and off at will, as he could modern colloquialism.

Another distinctive speaker is Glóin the dwarf, or perhaps one might say dwarves in general (though Glóin's son Gimli is the only named person present at the Council who does not speak, perhaps out of dwarvish deference to his father). Dwarves, it seems, like the Norsemen whose language Tolkien assigned to them, are characteristically taciturn. Glóin's sentences tend to be short, and to break where one might expect them to continue. He has a trick of using 'apposition', two constructions with much the same meaning, the second expanding the first: so, 'For a while we had news . . . messages reported that Moria had been entered'. He combines it sometimes with another trick found in both ancient and modern colloquial speech, which is to give several sentences which are formally parallel, but contain unspoken causal connections. See for instance the following, in which Glóin's connecting words are italicized, and the implied ones, which he leaves out, inserted in brackets:

'Also we crave the advice of Elrond. *For* the Shadow grows and draws nearer. [*As we can tell, because*] We discover that messengers have come also to King Brand in Dale, and that he is afraid. [*So*] We fear that he may yield. [*Because*] Already war is gathering on his eastern borders.'

At all times, furthermore, Glóin has a tendency toward oblique statements: when, interrupting Legolas's account of their merciful treatment of Gollum, he says 'You were less tender to me', he means of course not the opposite but the obverse, 'You were more cruel to me'.

Most striking – and suggesting that these are not characteristics of the idiolect of Glóin, but of the dwarves as a whole – is his account of the exchange with the messenger of Sauron. The messenger does not speak like a dwarf. For one thing he has an un-dwarvish tendency to say things three times: 'As a *small* token of your friendship Sauron asks this . . . that you should find this thief . . . and get from him, willing or no, a *little* ring, the *least* of rings, that once he stole.' Glóin stresses that he is quoting directly by adding to the word 'thief', 'such was his word', and later on parodies the messenger's speech-pattern by repeating mockingly the phrase 'this ring, this least of rings'. However, the messenger seems aware of the difference too, and at one moment tries to answer Dáin after his own manner, in the little interchange when he demands an answer:

> 'Dáin said: 'I say neither yea nor nay. I must consider this message and what it means under its fair cloak.'
> 'Consider well, but not too long,' [said the messenger].
> 'The time of my thought is my own to spend,' answered Dáin.
> 'For the present,' said he, and rode into the darkness.'

No one here means exactly what they say. Dáin's first statement is a veiled accusation ('fair cloak' implies foulness underneath). The reply looks like an agreement, but is a threat. Dáin's second statement seems to be so general, and metaphorical, as to be unanswerable. But the second reply, while again an agreement, qualifies that agreement so sharply as to be a threat again – 'there will be a time when your thoughts will not be free'. The exchange is menacing, it adds a touch of urgency to the question for which the Council has been called – 'what are we to do with the Ring?' – but it also creates strong characterization for the whole dwarvish race: stubborn, secretive, concealing their intentions, in a word (a word Tolkien uses, see Richard Blackwelder's *Tolkien Thesaurus*), 'thrawn'. It is appropriate that the Northern dialect word should be apparently the same as or close to the name of Thorin's father, Thráin.

Further revealing speech differences can be seen, for instance, in the contrast between Aragorn and Boromir: the only two Men at the Council, similarly imposing, with similar names, and indeed sharing common descent. They ought to talk the same way, but they do not, quite. Boromir's speech is from the beginning relatively Elrondian, one might say – he uses words like 'verily' and 'deem', and some inverted constructions, 'Loth was my father to give me leave'. Aragorn is capable of replying the same way, but tends to do it most when he is speaking directly to Boromir, as if to impress him. He is also capable of speaking quite colloquially, as in his reference to Butterbur – Boromir does not say things like 'one fat man', and shows no sign of knowing anyone of Butterbur's social status. There is of course a certain competition between the two, for if Aragorn were to be what he said, then he would displace Boromir from his position as Steward-to-be: Boromir never directly answers Aragorn's question, 'Do you wish for the House of Elendil to return to the Land of Gondor?' The clash of styles shows up when Boromir (for the second time) expresses doubt about what he has been told:

> 'Mayhap the Sword-that-was-Broken may still stem the tide
> – if the hand that wields it has inherited not an heirloom
> only, but the sinews of the Kings of Men.'

The doubt is potentially insulting, but Aragorn responds to it easily, almost chattily:

> 'Who can tell? . . . But we will put it to the test one day.'

However, though this is said easily, it contains within it a heroic formula often found in Old English ('now is the time', the heroes cry to each other, 'to put our boasts to the test'). Thorin Oakenshield says it too, in *The Hobbit*, though there Bilbo immediately and crossly mocks it, see pp. 42–3 above. Boromir replies to Aragorn ambiguously:

> 'May the day not be too long delayed.'

The way they talk reminds us, in miniature, that Aragorn is also Strider, and does not need to be on his dignity all the time; but at the same time that Strider is also Aragorn, and can claim just as much, indeed even more authority than Boromir. There is a hint of future trouble in the veiled challenges from both sides.

It is Gandalf's long monologue, however, which shows most variety in its use of 'impacted speakers', the direct speech of others quoted by Gandalf. Without that variety the immense amount of necessary plot-detail conveyed by the monologue would run flat. Several of Gandalf's (seven) 'impacted speakers' create, like Boromir or Sauron's messenger, a sense of the ominous, more or less concealed. Perhaps the least significant, in terms of plot, is Gaffer Gamgee, whose job is only to tell Gandalf that Frodo and the others have left. He makes too much of this, as Gandalf says, 'Many words and few to the point', and Gandalf stresses what it is he actually says:

'I can't abide changes, not at my time of life, and least of all changes for the worst.' 'Changes for the worst,' he repeated many times.

This is of course stupid, when all he has to complain about is the Sackville-Bagginses. Sharkey/Saruman will be much worse than this 'worst', and there could be yet worse than Sharkey. In any case the Gaffer simply does not know the meaning of his own words. When he says 'abide' it has the meaning, in context, of 'bear, put up with', and in fact the Gaffer can abide these – he just has, having no choice. In older use, though, the word means 'await the issue of, wait (stoically) for, live to see', and this would be a better use of it. The Gaffer does not learn, though, and at the end is still moralizing inaccurately. ' "It's an ill wind as blows nobody any good", as I always say' (but he didn't), 'And All's well as ends Better' (at least he has stopped using superlatives). The Gaffer is not very important, but he is a reminder of psychological unpreparedness; it may be remembered that Tolkien was writing 'The Council of Elrond' in the early years of World War II.

Isildur's scroll, meanwhile, which Gandalf has discovered in the archives of Gondor, and which tells how Isildur cut the Ring from Sauron's hand after the Battle of Dagorlad long ago, is even more archaic than Elrond's speech, using the old -eth verb-endings – 'seemeth', 'fadeth' – and subjunctive forms like 'were the gold made hot again'. Its most ominous feature, though, is timeless: Isildur says of the Ring, 'It is precious to me, though I buy it with great pain', and any reader of *The Hobbit* will remember that Gollum too called the Ring 'my precious'. Isildur in Gondor is already on the road to becoming a wraith – a fate from which Gollum has only just been saved. But the most ominous speaker in the whole chapter is also the most modernistic, and in a way the most familiar. It is Saruman, the wizard who has changed sides. Or has he? He wants to 'join with' the 'new Power', which is Sauron, for no other reason than that it is going to win. But when Gandalf shows no leaning towards that course, Saruman shows that he is prepared to betray the 'new Power' too. 'If we could command [the Ring], then the Power would pass to *us*. That is in truth why I brought you here.' If that is the 'truth', then why did Saruman make the other suggestion first? Because he doubted Gandalf's ambition, maybe? Or was there some truth in the argument he also put forward, that 'the Wise, such as you and I', might be able to persuade, direct, control Sauron – though he does not say Sauron, he says 'the Power'. The idea of anyone, however wise, persuading Sauron, would sound simply silly if it were said in so many words. No sillier, though, than the repeated conviction of many British intellectuals before and after this time that they could somehow get along with Stalin, or with Hitler.

Saruman, indeed, talks exactly like too many politicians. It is impossible to work out exactly what he means because of the abstract nature of his speech; in the end it is doubtful whether he understands himself. His message is in any case one of compromise and calculation:

'We can bide our time, we can keep our thoughts in our hearts, deploring maybe evils done by the way, but approving the high and ultimate purpose: Knowledge, Rule, Order; all things that we have so far striven in vain to accomplish, hindered rather than helped by our weak or idle friends. There need not be, there would not be, any real change in our designs, only in our means.'

The end justifies the means, in other words, a sentiment the twentieth century has learned to be wary of. One might note also the echo of Gaffer Gamgee at the start; and the unusual rhetorical polish of the balanced phrases, 'can bide / can keep, deploring . . . but approving, hindered rather than helped, need not be / would not be'. But it is all nonsense, summed up by the word 'real' at the end. What does 'real' mean when Saruman says 'real change'? The intention is clear enough, and one often hears people say things like that, but I do not think there is any logical answer to the question. When people say things like 'no real change', they mean there is going to be a major change, but they would like you to pretend it is minor; and too often we do. Saruman is the most contemporary figure in Middle-earth, both politically and linguistically. He is on the road to 'doublethink' (which Orwell was to invent, or describe, at almost exactly the same time).

The gist of the paragraphs above is only this. People draw information not only from what is said, but from how it is said. The continuous variations of language within this complex chapter tell us almost subliminally how reliable characters are, how old they are, how self-assured they are, how mistaken they are, what kind of person they are. All this is as vital as the direct information conveyed, not least, as has been said, to prevent the whole chapter from degenerating into the minutes of a committee meeting, which in a sense is what it is. Tolkien's linguistic control (a professional skill for him) is one of his least-appreciated abilities; there is a sour irony in observing critics with no linguistic knowledge presuming to tell him how to do it, or assuming it is some sort of accident. Nevertheless it should also be noted that 'The Council of Elrond' does one other thing, which is pass on information, and do it from an almost bewildering complex of directions.

The Council of Elrond: organizing the plot

If the Council were a well-organized committee meeting of modern times, it would have on its agenda only three items:

(1) to determine whether Frodo's ring is indeed the One Ring, the Ruling Ring
(2) if it is, to decide what action should be taken
(3) and further, who should take it.

All these issues are indeed stated explicitly during the Council. Elrond asks near the start, 'What shall we do with the Ring?' But more than twenty pages later Gandalf has to ask the question again, saying perfectly correctly, 'But we have not yet come any nearer to our purpose. What shall we do with it?' One reason the Council has not made any progress is that its members have been pre-occupied with issue (1) above. It is once again framed explicitly by Boromir, though only after the Council has been in session for some time, 'How do the Wise know that this ring is [Isildur's]?', and repeated by Galdor, 'The Wise may have good reason to believe that the halfling's trove is indeed the Great Ring ... But may we not hear the proofs?' The only conclusive proof is in fact this. The One Ring is known to have passed to Isildur, who cut it from Sauron's hand (and the Council has an eye-witness to this present in Elrond). Gandalf has seen a document written by Isildur himself which gives the inscription on the Ring he took; and Gandalf knows that that inscription is on Frodo's ring, because he tested it himself by throwing the ring in the fire and then reading the inscription on it, at Bag End in the Shire on April 13th 3018 – the Council itself takes place on October 25th of the same year. If Gandalf had said this at the start, the Council might have proceeded a good deal quicker. Why does this not happen? Is Elrond simply a poor committee chairman?

The way 'The Council of Elrond' is in fact presented can be summarized like this. The bulk of it – up to the point where Gandalf says, 'Well, the Tale is now told, from first to last', and then repeats the question, 'What shall we do with it?' – consists of some seven major accounts by different speakers, which I number to keep track of them. (1) is the dwarf Glóin's. This makes it clear that Sauron *thinks* that a ring connected with hobbits is of vital importance. Elrond's account (2), takes us much further back, to say what the Ring was in the beginning, and to trace it as far as Isildur and the ruin of the Gladden Fields. Logically, we should now hear about Gollum, who as Gandalf knows, and as he explained to Frodo six months before in the study at Bag End, took the ring from his friend Déagol, who found it in the Great River by the Gladden Fields – a vital point of connection. However, what happens instead is that concentration on the Ring is broken by Elrond's account of Gondor, and Boromir's response to it. The rhyme Boromir and his brother heard in their dream does contain the line 'For Isildur's Bane shall waken' (the Bane is the Ring), but it starts with 'Seek for the Sword that was broken', and the Council is distracted by having that explained, by being told about Aragorn's lineage, and by the slightly competitive by-play between Aragorn and Boromir. It is Boromir who brings them back to the point by his question quoted above, 'How do the Wise know that this ring is [Isildur's]?'.

He is answered first by Bilbo's résumé of events in *The Hobbit* (3) (which a wise chairman might well have cut short), and then by Frodo's account (4) of 'his dealings with the Ring from the day that it passed into his keeping', i.e. from September 22nd,

3001, to the date of the Council seventeen years later. While Frodo's narrative (4) is continuous from Bilbo's (3), though, Bilbo's is nothing like continuous from Elrond's (2). In fact there is a gap of almost three thousand years between them, from the death of Isildur in year 2 of the Third Age to the events of *The Hobbit* in year 2941. The gap can only be filled by the story of Gollum, and after Frodo has finished Galdor does indeed tenaciously repeat that no one has yet proved the identity of the rings in question. However he also raises the issue, where is Saruman? And Elrond calls on Gandalf to answer because 'the questions that you ask, Galdor, are bound together'.

This is not strictly true, and Elrond as chairman once again seems to be letting the meeting get out of hand. Gandalf's tale (5) begins slightly combatively by suggesting that Galdor cannot have been listening to the proceedings so far (something which often happens in committee meetings): do accounts (1) and (4) not show that Sauron at any rate thinks the ring is the Ring? Galdor might have replied, 'yes, but that's not proof', and Gandalf indeed concedes that there is 'a wide waste of time between the River and the Mountain', i.e. between Isildur losing the Ring in the River Anduin and Bilbo finding it under the Misty Mountains. But he still does not fill the gap decisively with what he knows about Gollum, or about the inscription on the Ring, going back instead to the year of *The Hobbit*, 2941 (when he failed to recognize Bilbo's ring for what it was), then to the year of Bilbo's departure from the Shire, 3001, and only then telling the story of his discovery of Isildur's scroll, with the vital identifying inscription on it. The sensible thing might then have been to go straight to the Shire and investigate Frodo's ring itself, but Gandalf makes a detour to talk to Aragorn, and Aragorn now gives his account (6) of the capture of Gollum. Only after that does Gandalf finally clinch the point about identity: he knows Frodo's ring is the Ring because he has seen the inscription on it, in the Black Speech, which he repeats. He also adds the fact, which now explains Glóin's account (1), that Sauron heard about the rediscovery of the Ring from Gollum; that is what started his enquiries about dwarves and hobbits. The Council is then side-tracked by Boromir's question about Gollum and Legolas's account (7) of his escape; and further by Gandalf returning to 'Galdor's other questions' (really only one question), 'What of Saruman?' Once Gandalf has told of his imprisonment in and escape from Orthanc, all he has left to do is to bring the story up to date by retracing his steps to the Shire, and following Frodo and party belatedly to Rivendell, the point at which he says, 'Well, the Tale is now told, from first to last'.

But to say that is to disguise a good deal of art. The tale has not in fact been told 'from first to last' at all, but through a series of interjections, as one character or another manages to turn the conversation to their immediate concerns. Nor has it been told in strictly logical order, with concentration on the central point, is the ring the Ring? Instead there have been distractions over Gondor, over Aragorn and the Sword that was broken, over Gollum, and over Saruman – all elements that will be vital to the plot, but are not directly vital to the Ring. And there is also a giant gap in the whole narrative, as said above nearly three thousand years, from year 2 to year 2941, a gap which Gandalf has filled for Frodo many chapters before with the tale of Sméagol and Déagol and the revelation that Gollum was once a hobbit, but which is left blank to the Council. It is a mark of Tolkien's skill in handling conversation that these gaps, loops, and meanders do not seem tedious, and indeed are usually not noticed. But then, as often happens in committee meetings, the real issue appears after much talk: what to do next?

Tolkien deals with this briskly, using three minor characters to put forward unsatisfactory answers before the true issue can be faced. Erestor suggests, give it to Bombadil: this is rejected. Galdor asks if it could simply be kept safe: this too is rejected. Glorfindel suggests, throw it in the sea: too dangerous, and not final enough. It is Elrond who says in the end, 'We must send the Ring to the Fire'. Over Boromir's objections he refuses to take it himself, and is backed up by Gandalf. The impasse, and the growing tension (and weariness) are relieved by a sudden drop in stylistic level, as the hobbits speak up, one after another, for almost the first time in this chapter in direct speech. Bilbo volunteers, commenting with near-bathos, 'It is a frightful nuisance', but putting his finger on the point – for (again as often happens in committees), it may be decided what is to be done without deciding who is to do it. And yet, as Bilbo says, 'That seems to me what this Council has to decide, and all it has to decide'. We have at last reached agenda item (3). At this moment Frodo speaks, with extreme plainness: 'I will take the Ring ... though I do not know the way'; and Sam breaks in for the first time, his presence till then not even mentioned: 'A nice pickle we have landed ourselves in, Mr Frodo!'

Of course, this decision is absolutely vital for the plot, the theme, and everything else. It is a testimony to Tolkien's art that he makes it seem a surprise, and also that the low level of hobbit style as the decision is made, with its chronic understatements ('frightful nuisance ... nice pickle'), remains dignified, even challenging in its plainness. However the final point to be made about 'The Council of Elrond' is this. It took Tolkien a very long time to reach this level of complexity – early drafts are given in *The Return of the Shadow*; and the first of them skips over almost all the matter discussed here in seven lines on p. 395. But once he had reached it, rather as with the arrival of the ring and of Gollum in *The Hobbit*, Tolkien had *solved his problem with story*. After this chapter there is a driving narrative requirement: take the Ring to Orodruin and destroy it. There is a series of loose ends: Gollum escaped, Saruman changing sides, Gondor under threat, Aragorn's status revealed but not acknowledged, the Three Rings of the elves, even the Mines of Moria and the indeterminate loyalties of Rohan. As the anonymous poet said of King Arthur and his knights, in the *Sir Gawain* poem Tolkien had edited nearly twenty years before, and in the words of Tolkien's own translation:

> Though such words were wanting when they went to table, now of fell work to full grasp filled were their hands.

More re-creations

Sir Gawain further provides a good example of the way Tolkien tended to work from this point on. He was now committed to taking up the challenge which *The Hobbit* had set: its careful creation of the sense that there was far more of Middle-earth than one story could ever bring into focus had naturally generated the desire for a sequel, and for more novelties in that sequel, and Tolkien now had to find them. He found them very largely where he had found the surprises of *The Hobbit*, in ancient literature – and in particular (and here very few of his imitators have been able to follow him) in the gaps and even the errors of ancient literature.

To pursue the example from *Sir Gawain*, in that poem the hero, rather like Frodo and company setting out from Rivendell, finds himself beset by all manner of dangers and hardships before ever he gets to the real goal of his quest. As the poet says (doing exactly what Tolkien had done in his account of Bilbo's return):

> Somwhyle wyth wormeȝ he werreȝ, and with wolues als,
> Sumqhyle wyth wodwos, þat woned in þe knarreȝ,
> Boþe wyth bulleȝ and bereȝ, and boreȝ oþerquyle,
> And etayneȝ, þat hym aneled of þe heȝe felle.

It is essential to see the original here, but Tolkien's translation of the lines runs:

At whiles with worms he wars, and with wolves also,
at whiles with wood-trolls that wandered in the crags,
and with bulls and with bears and boars too, at times;
and with ogres that hounded him from the heights of the
    fells.

The third word in the second line of the original presented a problem to editors like Tolkien and his colleague Gordon. It looks like a plural, indeed it must be a plural, to match all the other creatures mentioned in the four lines. But in that case, what is the singular of it? Presumably, *wodwo (another guess, or reconstruction). Tolkien and Gordon, however, did not like *wodwo. It did not have a sensible etymology. They concluded, instead, that the origin of the poem's plural form wodwos was in fact Old English *wudu-wása, a singular form, the plural of which would have been *wudu-wásan. This compound word, one might note, would then have been absolutely identical in form to Tolkien's later invention *hol-bytla, plural form (several of Tolkien's critics have got this wrong, including C.S. Lewis) *hol-bytlan. In both cases the first element of the compound is common and familiar, being no more than the ordinary words for 'wood' and 'hole', but the second element is rare or unknown: and the compound designates a non-human creature to be guessed at only from the word. Tolkien thought, in brief, that the Gawain-poet should have written wod-wosen, not wodwos; but he or the copyist had assumed, wrongly, that the -s at the end of the rare word wodwos was a plural already.

Having gone so far, though, the next question (a natural one, but one not to be answered by philology) was, 'what, then, are these wood-woses?' Tolkien answered the question in Book V/5, of The Lord of the Rings, where we meet, for a moment, 'the Woses, the Wild Men of the Woods'. One other element in his thinking at that stage, though, may well have been this. His office at the University of Leeds was just off a road called Woodhouse

Lane, down which he had to come every day from his house in Darnley Road. Woodhouse Lane leads over Woodhouse Ridge and Woodhouse Moor, the latter areas still wooded and largely undeveloped even now because of the steep fall down to the stream at the bottom. Of course, 'woodhouse' in these names could just have the dull meaning, 'house in the wood'. On the other hand, Tolkien thought that the modern surname Woodhouse derived from *wudu-wása, and he knew also that in several Northern dialects, 'wood-house' and 'wood-wose' would be pronounced exactly the same, i.e. 'wood-'ose'. So the modern spelling of Woodhouse Lane could be a mistake, just like the *Gawain*-poet's. The road he went up and down every working day might preserve, in a completely prosaic context, a memory of uncanny creatures, the 'wild men of the woods' who once haunted the tangles above the River Aire. And if *wudu-wásan could survive misunderstood and worn down as 'woodhouses', one might add, then why should *hol-bytlan not survive as 'hobbits'?

Tolkien's creation of the Woses shows at once his dependence on ancient texts, his conviction that at times he knew better than the authors (or anyway than the copiers) even of those texts, and his ability to set academic puzzles in entirely contemporary contexts. His inventions often sprang from words, or from names. But in investigating the words, and the names, he worked on the principle that they must at one time have had known referents, which with patience and imagination could be recovered. Middle-earth was always to him, as suggested in the 'Foreword', an 'asterisk-reality': it had not been recorded, like the *-forms of early words, but again like the *-forms it could be inferred, or reconstructed, with high plausibility if not complete certainty. The guarantee of Middle-earth, as of the verbal reconstructions of philologists, was inner consistency. The Woses are not demanded by the plot of *The Lord of the Rings*, but they feel as if they should be there. They help to create the fullness which is the major charm of Tolkien's Middle-earth.

Tolkien used what one might call the 'wood-wose' model of invention increasingly from the end of 'The Council of Elrond' and the departure of the Fellowship. The scenes on Caradhras and in Moria, one might say, tend to recapitulate early chapters of *The Hobbit* – in both cases a company setting out from Rivendell. The storm on Caradhras is rather like the storm in the Misty Mountains, with both times a feeling that the storm is not just a natural phenomenon; the 'wild howls of laughter' and the 'fell voices on the air' which Boromir recognizes echo the 'stone-giants . . . hurling rocks at one another for a game' and 'guffawing and shouting all over the mountainside' in chapter 3 of *The Hobbit*. In the same way, the entrance into Moria is rather like the entrance into the goblin-tunnels in *The Hobbit*, with much the same outcome – adventures in the dark leading to a passage to the other side of the mountains. A new element in Moria, though, is the Balrog, introduced in exactly the same way as so many of Tolkien's inventions, as if we ought to have known about it already: 'A Balrog', muttered Gandalf. 'Now I understand.' But we do not.

Just like Woses, Balrogs owe a part of their existence, at least, to an editorial problem. There is an Old English poem called *Exodus*, like several Old English poems a paraphrase of a part of the Bible. Tolkien's edition of it did not appear during his lifetime, but it came out posthumously, appropriately enough 'reconstructed' from his lecture notes. Since it is both a paraphrase and a fragment, the poem has never managed to gain a central position in literary courses, but Tolkien was interested in it: for one thing, he thought on linguistic grounds that it was older than *Beowulf*, and he thought that like the *Beowulf*-poet, the *Exodus*-poet had known a good deal about the native pre-Christian mythology, which could with care be retrieved from his copyists' ignorant errors. In particular, the poet at several points mentions the *Sigelwara land*, the 'land of the Sigelware'. In modern dictionaries and editions, these 'Sigelware' are invariably translated as

'Ethiopians'. Tolkien thought, as often, that that was a mistake. He thought that the name was another compound, exactly like *wudu-wása* and *hol-bytla*, and that it should have been written *sigel-hearwa*. Furthermore, he suggested (in two long articles written early on in his career, and now ignored by scholarship) that a *sigel-hearwa* was a kind of fire-giant. The first element in the compound meant both 'sun' and 'jewel'; the second was related to Latin *carbo*, 'soot'. When an Anglo-Saxon of the preliterate Dark Age said *sigelhearwa*, before any Englishman had ever heard of Ethiopia or of the Book of Exodus, Tolkien believed that what he meant was 'rather the sons of Múspell [the Old Norse fire-giant who will bring on Ragnarök] than of Ham, the ancestors of the Silhearwan with red-hot eyes that emitted sparks, with faces black as soot'.

The fusion of 'sun' and 'jewel' perhaps had something to do with Tolkien's concept of the *silmaril*. The idea of a fire-spirit re-emerges in the brief glimpse of the orc-chieftain who stabs Frodo, with his 'swart' face, red tongue and 'eyes like coals', but it also gave Tolkien Durin's Bane, the Balrog. Tracing such an ancestry of course says nothing about the way the Balrog is deployed, or the increasing tension of the Moria chapters, 'A Journey in the Dark' and 'The Bridge of Khazad-dûm'. One apparent aspect of these, however, is their strong antiquarianism – the interest in elvish runes in the inscription on the West-Door, the dwarvish runes on the tomb of Balin (both of them reproduced in full), the image of Gandalf poring over a tattered manuscript in the Chamber of Mazarbul. Another is their relative understatement. Unlike many of his imitators, Tolkien had realized that tension was dissipated by constant thrill-creation. Accordingly the dangers of Moria build up slowly: from the first reluctance of Aragorn, 'the memory is very evil' (never enlarged on), to the ominous knocking from the deep that answers Pippin's stone (was it a hammer, as Gimli says? – we never learn), to Gandalf's mention of Durin's Bane. The Balrog is also hinted at several times before it appears: the orcs hang back as if they are afraid of something on their own side, Gandalf contests with it and concedes 'I have met my match' before it is ever seen, and again the orcs and trolls fall back as it comes up to cross the bridge of Khazad-dûm. Even when it does come into focus, the focus is blurred:

> Something was coming up behind them. What it was could not be seen: it was like a great shadow, in the middle of which was a dark form, of man-shape maybe, yet greater; and a power and a terror seemed to be in it and to go before it.

The clash of Gandalf and the Balrog produces yet further feelings of mystery: we hear of, but do not understand, the opposition between 'the Secret Fire . . . the flame of Anor', and 'the dark fire . . . flame of Udûn'. What Tolkien does in such passages is to satisfy the urge to know more (the urge he himself felt as an editor of texts so often infuriatingly incomplete), while retaining and even intensifying the counterbalancing pleasure of seeming always on the edge of further discovery, looking into a world that seems far fuller than the little at present known. If gold and greed and mastery are 'the desire of the hearts of dwarves', then words and links and inferences are the lust of philologists. Tolkien had that lust as strongly as anyone ever has, but he felt it was one which could be strongly shared.

As with 'The Council of Elrond', a full survey of all Tolkien's philological roots would be too exhaustive to follow readily, but three final examples of his methods may be briefly given. First, orcs. Tolkien had used the word in *The Hobbit*, but his regular word at that point was 'goblin'. As he built up the linguistic correspondences of Middle-earth, mentioned above, this came to seem out of place: it is a relatively late word in English (the *OED* cannot find a clear citation before the sixteenth century), and according to the *OED* it derives probably from medieval Latin *cobalus* – the dictionary oddly makes no attempt to link it with the German-derived 'kobold'. Tolkien preferred an Old English word, and found it in two compounds, the plural form *orc-neas* found in *Beowulf*, where it seems to mean 'demon-corpses', and the singular *orc-þyrs*, where the second half is found also in Old Norse and means something like 'giant'. Demons, giants, zombies – it seems that literate Anglo-Saxons really had very little idea what orcs were at all; but then they or their descendants had the same problem with woses and *sigelhearwan*. The word was floating freely, with ominous suggestions but no clear referent. Tolkien took the word, brought the concept into clear focus in detailed scenes (to be discussed in the next chapter), and, as with hobbits, has in a way made both word and thing now canonical.

A parallel case is that of the Ents. 'Ent' is another Old English word, found relatively frequently in that language but even more puzzling than 'orc' or 'wose'. Its first association seems to be one of gigantic size: dictionaries usually translate *ent* as 'giant', and the huge sword which Beowulf snatches up in Grendel's mother's lair is the work of ents, *enta ærgeweorc*. But ents also seem to have had a reputation as builders in stone. Anglo-Saxons faced with Roman roads and Roman ruins were liable to describe them as *orþanc enta geweorc*, 'the cunning work of giants'. Tolkien perhaps read the first word not as an adjective but as a name, so that the phrase now means 'Orthanc, the ents' fortress' – which is what Tolkien's Orthanc in the end becomes and might seem to later human generations always to have been. However the main point about Anglo-Saxon ents is that whatever they were, they are felt, unlike ettins or thurses or elves or dwarves, to be no longer present, no longer a threat: they are present only in their surviving artefacts. Tolkien picked up the gigantism, the connection with the word or name Orthanc, and most of all the sense of extinction, for that is the fate of the ents even more than of the other non-human peoples of Middle-earth. What came entirely from his own mind was the connection with trees, the idea of ents as tree-herders, creatures which rise from and turn back into wood, as trolls do with stone: a part of what in later years would come to be seen as Tolkien's 'Green' ideology.

A final example may be Lothlórien. The 'magic' of Lothlórien has many roots (some of them to be discussed later on), but there is one thing about it which is again highly traditional, but also in a way a strong re-interpretation and rationalization of tradition. There are many references to elves in Old English and Old Norse and Middle English, and indeed in modern English – belief in them seems to have lasted longer than is the case with any of the other non-human races of early native mythology – but one story which remains strongly consistent is the story about the mortal going into Elfland, best known, perhaps, from the ballads of 'Thomas the Rhymer'. The mortal enters, spends what seems to be a night, or three nights, in music and dancing. But when he comes out and returns home he is a stranger, everyone he once knew is dead, there is only a dim memory of the man once lost in Elf-hill. Elvish time, it seems, flows far slower than human time. Or is it far quicker? For there is another motif connected with elves, which is that when their music plays, everything outside stands still. In the Danish ballad of 'Elf-hill' (*Elverhøj*), when the elf-maiden sings: 'The swift stream then stood still, that before had been running; the little fish that swam in it played their fins in time'. Tolkien did not at all mind deciding that ancient scribes had got a word wrong, and correcting it for them, but he was at the same time reluctant ever to think that they had got the whole story wrong, just because it did not seem to make sense: it was his job to make it make sense. Lothlórien in a way reconciles the two motifs of the 'The Night that Lasts a Century' and 'The Stream that Stood Still'. The Fellowship 'remained some days in Lothlórien, as far as they could tell or remember'. But when they come out Sam looks up at the moon, and is puzzled:

'The Moon's the same in the Shire and in Wilderland, or it ought to be. But either it's out of its running, or I'm all wrong in my reckoning.'

He concludes, it is 'as if we had never stayed no time in the Elvish country ... Anyone would think that time did not count in there!' Frodo agrees with him, and suggests that in Lothlórien they had entered a world beyond time. But Legolas the elf offers a deeper explanation, not from the human point of view but from the elvish (which no ancient text had ever tried to penetrate). For the elves, he says:

'the world moves, and it moves both very swift and very slow. Swift, because they themselves change little, and all else fleets by: it is a grief to them. Slow, because they do not count the running years, not for themselves. The passing seasons are but ripples ever repeated in the long long stream.'

What Legolas says makes perfect sense, from the viewpoint of an immortal. It also explains how mortals are deceived when they enter into elvish time, and can interpret it as either fast or slow. All the stories about elves were correct. Their contradictions can be put together to create a deeper and more unpredictable image of Elfland, at once completely original and solidly traditional.

Cultural parallels: the Riders of the Mark

At the end of The Fellowship of the Ring the story seems to open out, and also to start gathering pace. First Celeborn gives the Fellowship a verbal map of what lies before them – 'There the Entwash flows in by many mouths from the Forest of Fangorn in the west' – and then Aragorn follows suit, 'You are now looking south-west across the north plains of the Riddermark, Rohan of the Horse-lords'. In quick succession then, in the first Book of *The Two Towers*, we are introduced to the Riders, the Uruk-hai, the Ents, and Saruman; one might note also that in Tolkien's own careful chronology of events in Appendix B, the action of Book I (Bilbo's farewell party to Frodo's arrival in Rivendell) occupies seventeen years, the action of Book II three months (November 20th 3018 to February 26th 3019), but that of Book III only ten days, Book IV only fifteen. The sense of increasing speed, however, may also have something to do with the arrival on the scene of the Riders of Rohan. Even at a late stage of composition Tolkien had no idea that they were going to come in, but when they did, one imagines that matters immediately became easier for him: when Tolkien wrote about the Riders he had plenty of material ready to hand.

The fact is that the Riders, like the hobbits, are another image of Englishness – Old English, of course, not modern English, but English just the same. Tolkien later on denied this, insisting in a footnote to Appendix F (II) that the fact that he had 'translated' all Rider-names into Old English did not mean that Riders and Anglo-Saxons were any more than generally similar. But the 'translation' process runs very deep. One might begin with the Riders' word for their own country (as translated), 'the Mark'. This is, or ought to be, almost as familiar as 'the Shire', but it has been obscured by just the kind of learned Latinizing (and Frenchifying) which Tolkien so much disliked and defied in names like 'Bag End'. Among historians the central kingdom of Anglo-Saxon England is invariably known as 'Mercia', its inhabitants as 'Mercians'. These must however be Latinizations of native terms, and indeed the West Saxons (in whose dialect most Old English texts survive) called their neighbours the *Myrce*. If their name for their neighbours' kingdom had survived it would certainly have been, the **Mearc*. Tolkien, however, a native of Mercia, as he often proclaimed, would have had no trouble in translating this back into Mercian, removing the West Saxon diphthongization and coming up with the *Marc, pronounced 'Mark'. In Anglo-Saxon times the natives of Worcestershire and Warwickshire and Oxfordshire, and many other counties, would have told anyone who asked that they lived in the Mark, and also in their own particular Shire: names at once both ancient and modern, indeed unchanging. As for the white horse that is the emblem of the Mark, like Bree and the Barrow-downs it lies less than a day's walk from Tolkien's study, the White Horse of Uffington, cut into the chalk a short stroll from the great Stone Age barrow of Wayland's Smithy. All the names given to the Riders and their horses and their weapons are pure Anglo-Saxon, and (a point less often noted) pure Mercian, not West Saxon. The names of their kings, Théoden, Thengel, Fengel, Folcwine, etc., are all simply Anglo-Saxon words or epithets for 'king', except, significantly, the first: Eorl, the name of the ancestor of the royal line, just means 'earl', or in very Old English, 'warrior'. It dates back to a time before kings were invented.

There is, one has to admit, one thing about the Riders which does not resemble the historical ancestors of the English, which is that they are riders. In texts of the later Anglo-Saxon period like the poem *The Battle of Maldon* or the prose *Anglo-Saxon Chronicle*, the reluctance of the Anglo-Saxon military to have anything to do with horses approaches the doomed, or the comical. The *Maldon* poem begins with the English leader telling his men to leave their horses and advance on foot; an attempt by an English army to fight on horseback in the year 1055 ended in total catastrophe, according to the *Chronicle* (which blames the whole fiasco on Norman leadership). It could be argued that Hastings was lost because of this insular insistence on fighting on foot. Nevertheless, Tolkien might have thought, the evidence, deeply considered, might not all point the one way. It is striking, for instance, that in modern English there is no native word for cavalry country, a flat sea of grass: we make do with foreign words like 'steppe', 'prairie', 'savannah'. The reason is obvious:

there are no steppes or prairies in England. If there were, though – and the likeliest place to look is on the flatlands of East Anglia – we might have had a word for them, and the word (Tolkien worked out) would be 'emnet'. This is in fact the first place name in the Riddermark that we are given, with the Eastemnet being mentioned early on in III/2, as Aragorn and the others pursue the Uruk-hai, the Westemnet by Éomer a few pages later. Emneth, though, is also a real place in Norfolk. It probably derives from Old English *emn-mæþ*, or in modern English 'even meadow', quite clearly the same idea as 'steppe' or 'prairie'. If the Old English *had* encountered a sea of grass, then, they would have called it an emnet, and it is possible to imagine what would have happened to them if they did. In any case, before they emigrated to the island of Britain, perhaps they had. Tolkien may well have known that the peculiar and rather unsatisfactory set of Old English words for colour – grey, dun, fallow etc. – could well be interpreted as words for the colour of horsehide. He uses one of them in the name of the horse lent to Aragorn, Hasufel, in modern English 'dark-skin'. The word Éomer uses early on in his order to Éothain, 'Tell the *éored* to assemble on the path', is another clue. It is a word of exactly the same class as *sigel-hearwa*, one never used in any text we have, but surmised by editors to explain a mistake in a line of poetry. Line 62 of the Old English poem *Maxims I* reads, *Eorl sceal on éos boge, worod sceal getrume rídan*, 'earl shall [go] on horse's back, warband (*worod*) ride in a troop'. The alliteration fails on *worod*, the word is spelled wrong, and in any case Anglo-Saxon warbands (see above) normally marched on foot. Editors solve the problem by crossing out *worod* and writing in the asterisk-word *éored*, 'mounted troop' – to be carefully inserted in exactly the right context by Tolkien as he started to create the image of the Riddermark. Maybe the ancestors of the English, therefore, were not as hippophobe as later records suggest; their descendants have after all been passionately hippophile ever since.

The Riders of the Mark are then a reconstruction from many sources, like so much in Tolkien, a blend of ancient and modern, the strange and the familiar, the learned (like **éored*) and the absolutely matter-of-fact (like the place-name Emneth). The underlying model for much of what they do and how they behave is furthermore perfectly obviously the Old English epic of *Beowulf*, which Tolkien knew so well. Théoden's hall is called Meduseld; so is Beowulf's. The courts round it are called Edoras; see again *Beowulf*, line 1037. In the chapter 'The King of the Golden Hall', the etiquette of arrival and reception corresponds to that of *Beowulf* point for point. In the epic Beowulf and his men are challenged by a Danish coastguard, who hears what they have to say, makes his own decision to let them pass, and escorts them to the hall of King Hrothgar itself. Here he leaves them, to be met again by a doorwarden, who keeps the visitors outside till he has gone in and reported their arrival to the king; he then comes back, invites them in, but tells them firmly to leave their weapons outside: 'Let the battle-shields wait here . . .' Beowulf then goes in, to be greeted by the king, but then, shortly after, to be challenged and indeed insulted by the king's counsellor, who 'sat at the feet of the lord of the Scyldings'.

All this is exactly what happens in *The Two Towers*. Gandalf, Aragorn and company are met by an outer guard, who passes them on to the door-guards, saying as he does so almost exactly what the Danish coastguard says in the poem – the one, 'I return to the sea, to hold guard against any fierce warband', the other, somewhat redundantly, 'I must return now to my duty at the gate'. The clash between Háma and Aragorn over relinquishing weapons then has no counterpart in *Beowulf* (Beowulf of course prefers like Beorn to fight bare-handed, and so has little concern for weapons). However, Tolkien has here revealingly transposed a scene of tension from early on in the sequence of events to rather later. In the epic, the critical moment was at the coastline, when the coastguard had to make up his mind about letting them pass on Beowulf's word alone. He thinks it over, and then decides, in words which have been disputed by editors and translators, but which are clearly a gnomic maxim of some kind:

> Æghwæþres sceal
> scearp scyldwiga gescad witan,
> worda ond worca.

My translation of them would be, 'A sharp shield-warrior must be able to decide, from words as well as from deeds' (for, unstated, any fool can decide from deeds – it is deciding *before* anything has happened which is the test of intelligence). Tolkien clearly pondered the saying, and decided to rephrase it without changing its point. After Háma has forced Aragorn, Legolas, Gimli and Gandalf to hand over their undeniable weapons, the question arises of Gandalf's staff. Is it a weapon or not? 'The staff in the hand of a wizard may be more than a prop for age', says Háma – and indeed we soon learn both that it is, and that Gríma Wormtongue had foreseen the point. But Háma decides to let it pass:

> 'in doubt a man of worth will trust in his own wisdom. I believe you are friends and folk of honour, who have no evil purpose. You may go in.'

His three sentences match those of the Danish coastguard point for point. The first one paraphrases the maxim given above. The second one is the same as the coastguard's 'I hear that this war-band is friendly', the third parallels his statement of permission, 'Go forward'. Háma's first sentence falls into the four-stress, alliterative beat of the Anglo-Saxon gnome (doubt / trust, worth / wisdom), for the Riders, like the Anglo-Saxons, prize sententious sayings: Aragorn uses one tactically on the court-guard, 'seldom does thief ride home to the stable'.

The interchanges with the guards however do more than merely show that Tolkien was copying *Beowulf*. They make a point about the Riders' sense of honour, and of proper behaviour, in a sense wider than mere formulas of greeting. At several places one can see that the Riders, while disciplined in a way, do not have the rigid codes of obedience of a modern army, or a modern bureaucracy. Éomer is not sure whether or not he should lend Aragorn, Legolas and Gimli horses. He knows that Théoden is already angry about the loss of Shadowfax; and he has received an order to take strangers to the king. Nevertheless, faced with Aragorn's direct refusal, and explanation of his quest, he decides to disobey: 'You may go; and what is more, I will lend you horses.' He knows this may lead to his own punishment, even execution – 'I place myself, and maybe my very life, in the keeping of your good faith' – but he does it just the same. One notes that his second-in-command Éothain immediately queries the order, though he gives in in the end. The unnamed guard of the court-yard decides to disobey an order as well, that 'no stranger should pass these gates', because he declines to take it from Wormtongue; and Háma then, as we have seen, uses his own judgement on the issue of the staff. He also gives Éomer his sword back before being told to do so, and Théoden notices the fact. In modern times such an anticipation of the findings of the court-martial, so to speak, would be a serious offence and would not be passed over. But Théoden lets it go. All these scenes make a point about freedom. The Riders are indeed 'a stern people, loyal to their lord', but they are not governed, as we are, by written codes. They are freer to make their own minds up, and regard this as a duty. They surrender less of their independence to their superiors than we do ; and Tolkien makes us realize that even if they are relatively 'uncivilized', indeed still at a 'barbarian' stage of development, this is not all bad. They can be at once more ceremonious and more relaxed than modern people.

The oral nature of the Riders' culture is further stressed in several ways. Almost the first thing Aragorn says about them is that they are 'wise but unlearned, writing no books but singing many songs'. Later on he gives an example of their poetry, 'Where now the horse and the rider? Where is the horn that was blowing?', based on the Old English poem of *The Wanderer*; Théoden's 'call to arms' in III/6, 'Arise now, arise, Riders of Théoden', is based on the poem *The Finnsburg Fragment*, Tolkien's commentary on which was published posthumously in 1982; in *The Return of the King* we have more extended Rider poetry, written in strict Old English metre, in the song of the ride to Gondor (V/3), the song of 'the Mounds of Mundburg' (V/6), and Gléowine's song in memory of his master Théoden (VI/6). Nearly all the poetry that is quoted is strongly elegiac, one might note: in a culture with no written records that is a major function of poetry, at once to express and to resist the sadness of oblivion. It has the same function as the spears that the Riders plant in memory of the fallen, as the mounds that they raise over them, as the flowers that grow on the mounds. Éomer says as he passes one burial-place, 'when their spears have rotted and rusted, long still may their mounds stand and guard the Fords of Isen!' As they ride up to Meduseld between the royal barrows (paralleled in reality by the burial-sites of Sutton Hoo in England and Gamle Lejre in Denmark), Gandalf looks at the white flowers that cover them and says, 'Evermind they are called, *simbelmynë* in this land of men, for they blossom in all the seasons of the year, and grow where dead men rest.' Tolkien makes a point here uncommon in the many attempts to present the barbarian past: that the very fragility of record in such societies makes memory all the more precious, its expressions both sadder and more triumphant. As often, his imaginative re-creation of the past adds to it an unusual emotional depth.

Cultural contrasts: Rohan and Gondor

There is one final point that may be made both about the Riders'
behaviour patterns, and the images which correlate with them,
but it is a point which has to be made comparatively. As has
been remarked before, Tolkien clearly meant to set up a contrast
between the Riddermark and Gondor, and he did it on several
levels, tacitly contrasting Théoden with Denethor, for instance (a
comparison to be discussed in the next chapter), and setting up
Boromir (in Éomer's opinion) as a sort of middle term: 'More
like to the swift sons of Eorl than to the grave Men of Gondor
he seemed to me, and likely to prove a great captain of his people
when his time came.' Éomer clearly thinks the Riders superior
in action to the Gondorians; but the Gondorians think that the
Riders are just 'Middle People', intermediate between the truly
civilized (themselves) and the 'Men of the Wild' like the Dunlend-
ings or the Haradrim. Who is right, what is the difference, is it
is possible that both are right? Tolkien answers these questions
silently in his narrative, deepening his presentation of the two
different cultures as he does so.

Two obviously contrasted scenes are the two in which first
Pippin and then Merry offer their service to Denethor and
Théoden respectively, in V/1 and V/2. The central action in each
case is the same: the hobbit offers his sword to the man, who
accepts it and returns it. The central similarity however only
points up the surrounding differences. Merry's action is spon-
taneous, prompted only by 'love for this old man', and is received
in the same spirit. The ceremony, in so far as there is one, consists
only of Merry saying 'Receive my service, if you will', and
Théoden replying 'Gladly will I take it ... Rise now, Meriadoc,
esquire of Rohan of the household of Meduseld'. There is no
doubt about the binding quality of what has happened, but it
takes few words. By contrast Pippin's offer has more complex

motives: pride, and anger at the 'scorn and suspicion' in Denethor's questioning. His offer is not immediately accepted:
Denethor looks at his sword first, the one taken from the Barrow-
wight, and seems to be affected by that before he says 'I accept
your service' (not quite the same words as Théoden's, for the
one uses the colloquial 'take', the other the formal 'accept'). Both
parties in the Minas Tirith scene then make a formal statement,
naming themselves in full and giving both patronymics and titles:
Denethor's is not without an element of threat, a promise to
reward 'oath-breaking with vengeance', far removed from Théoden's 'Take your sword and bear it unto good fortune'. It is
probably fair to say that the scene between Merry and Théoden
makes much the better impression, kindlier, more casual, and
with more concern for the feelings of the junior party.

One might say the same of the equally contrastive descriptions
of the two men's halls. Théoden's is shadowy, but pierced by
bright sunbeams. It has a mosaic floor and painted pillars. Its
most obvious feature, though, is the 'Many woven cloths' on the
walls, with the picture of Eorl the Young, done in green and
white, red and yellow, picked out by the sun. All these features
echo the hall of King Hrothgar in *Beowulf*, Heorot (like Théoden's
Meduseld and Beowulf's own Meduseld, doomed to be devoured
by fire); Heorot too has a 'painted floor', gilded gables, and 'webs'
or tapestries to adorn it on ceremonial occasions. Tolkien has
added only one foreign word to the description of Meduseld,
'louver', the device which allows smoke to escape from the wood-
burning but chimneyless building. The hall of Minas Tirith is
markedly different – also 'lit by deep windows' and held up by
many pillars, but quite without colour, or one might say without
life. 'No hangings nor storied webs, nor any things of woven stuff
or of wood, were to be seen in that long solemn hall', only 'a
silent company of tall images graven in cold stone': this time the
native word, 'web', i.e. tapestry, stands out as alien to Gondor in
the same way that 'louver' stood out as alien to the Mark. A

further difference is that where Théoden sat in 'a great gilded chair' on a dais with Wormtongue at his feet on the steps leading up to it, Denethor's hall has a dais and a throne, but the throne is empty, and Denethor is sitting, with a kind of humility, on a plain chair on the bottom step, in the same ceremonial space as Wormtongue: he is a steward, not a king. It is clear, though, that Denethor's humility masks an evident pride, as he shows in his rebuke of Gandalf, 'the rule of Gondor, my lord, is mine and no other man's, unless the king should come again'. His exchange with Gandalf in a way repeats in its tone the near-clash between Aragorn and Boromir in 'The Council of Elrond', the Gondorian striving for superior dignity, the other party asserting superior status, but feeling no need to mark this formally.

What right have the Gondorians, then, in their claim to be 'High' rather than 'Middle', as Faramir explains? One may see something of it in a third pair of contrasted scenes, the two encounter-scenes between, on the one hand, Éomer and Aragorn in the Eastemnet (III/2), and on the other, between Faramir and Frodo in Ithilien (IV/5). The similarities are close and frequent. In both cases an armed company comes upon strangers in a disputed borderland. In both cases the leader of the company is under orders to arrest strangers and take them back, but decides not to obey the order, at the risk of his own life. Both scenes begin with a hostile demonstration, indeed a surrounding, and in both a subordinate member of the weaker party (Gimli, Sam) comes close to losing his temper in support of his own leader. In both scenes, finally, there is an initial sequence which is public, heard by all the Riders or the Rangers, and then a second one in which the leader of the group speaks more privately and in more conciliatory fashion. However, the scenes make a quite different impression, and in this case, unlike the two just mentioned, the balance is on the whole in favour of Faramir and Gondor.

The first thing one might say about the Riders in the Eastemnet scene is that they contain an element more familiar from America than from England. Their sudden wheel and narrowing circle round the strangers, weapons poised, is more like the old movies' image of the Comanche or the Cheyenne than anything from English history. Aragorn's impassive response to it also suggests that he understands the stoicism ascribed by tradition to native Americans. Éomer's behaviour is then markedly aggressive, and the two sides all but come to blows – though he has some cause, for Aragorn and his fellows are in fact slow to give their true names, insisting on hearing his first, and insisting on answers to their questions before they answer his. It is possible to sympathize with Éomer's remark that 'wanderers in the Riddermark would be wise to be less haughty in these days of doubt'. Tension recedes once Éomer calls off his men, and begins to talk with less sense of his own dignity; he accepts doubt, even correction, is prepared to explain himself, comes close to apologizing. Nevertheless the general sense one has of him is that while he may indeed be at bottom able and good-natured, he is out of his depth in dealing with Aragorn, and moreover more 'swift' or impetuous than is quite safe. As usual there is an object which gives a visual focus for this feeling, and it is Éomer's main distinguishing feature: the white horsetail on the crest of his helmet. He is repeatedly picked out by this decoration. It is a nomad image, derived from the steppe-dwellers, the Turks or the Scythians (though one may well remember that the English have adopted it cheerfully, using horsetail plumes to decorate the helmets of Her Majesty's Life Guards to this day). There is even a word in English for both the plume and the quality it represents, though of course the word is foreign-derived: it is *panache*, which means at once the decoration in a knight's helmet, and the cavalryman's swagger, the sudden onset that sweeps away resistance.

If Éomer has panache, though, Faramir has something more valuable, a kind of prudence or restraint. He too is capable of a certain grimness or asperity when he is grilling Frodo, but when he is interrupted by an irate Sam he puts the interruption aside 'without anger'. He also keeps his cards much closer to his chest than does Éomer, concealing the fact that he has seen the body of his brother Boromir until Frodo has mentioned him several times. In addition, he notes Frodo's hesitation over his relations with Boromir, but jumps to the right rather than the wrong conclusion. Two striking things about the conversation, though, are first that Faramir (in direct contrast to Éomer) knows more about Lothlórien even than Frodo does, and corrects him about its effects rather than having to be corrected; and, most important, that he holds off his interrogation and deliberately chooses not to press an advantage once he realizes that Frodo is concealing something about 'Isildur's Bane'. By doing so, he leads Sam to blurt out the truth, which says something about the efficiency of Faramir's tactic; but he also sees further into the future than did Boromir, or Éomer. If Éomer is a pleasant young man, as said above, Faramir is a 'grave young man'. His rejection of mere militarism, his recognition that there are other qualities than those of a warrior or a general, backs up his claim that Gondor is a more reflective society, and one with a longer history, than the Riddermark. The claim is also tacitly demonstrated by Faramir's capacity for subtlety, understatement, a reverence for truth which nevertheless includes a relatively oblique approach to it, well beyond Éomer's blunt aggressions and withdrawals.

The ironies of interlace

The first three Books of *The Lord of the Rings* can be seen in particular as a kind of complex map, a map of cultures, races, languages, and histories, which gives the world through which the characters move its special depth and being. The map would remain powerful, and have a kind of charm, even if the characters were just strolling through it (as sometimes they are, early on).

However the contrastive element which comes in as the material for contrasts accumulates has a further analogue in the way the story is told. *The Two Towers* especially, and the first part of *The Return of the King*, have a structure reminiscent on a large scale of 'The Council of Elrond' on a small one. The word that describes the structure is 'interlace'.

Tolkien certainly knew the word, for it has become a commonplace of *Beowulf*-criticism, but he may not have liked it much: it is associated also with the structure of French prose romance, in which he took little interest. However, Tolkien certainly also knew that the Icelandic word for a short story is a *þáttr*, literally a thread. One could say that several *þættir*, or threads, twisted round each other, make up a saga; and Gandalf comes close to saying something like that when he says to Théoden, 'There are children in your land who, out of *the twisted threads of story*, could pick the answer to your question' (my emphasis). Tolkien may have felt that there had been all along a native version of the French technique of *entrelacement*, even if we no longer know the native word for it. But word, or no word, he was going to do it.

One can see this by considering the way the narrative is arranged from the start of *The Two Towers* to chapter 4 of Book VI of *The Return of the King*, 'The Field of Cormallen'. I have attempted to represent this by the accompanying diagram. It should be noted that the diagram is in several ways a simplified one. It does not show the brief separation of Legolas and Gimli during IV/7; full representation of the movement of characters on 15th March would require a separate diagram; and there are of course many characters involved in the story besides the members of the Fellowship. Nevertheless the diagram may illustrate the nature of the narrative threads and their 'interlacing'.

The diagram covers the period from the start of *The Two Towers* (February 26th) to Book VI, chapter 4, of *The Return of the King* (March 25th). During this period, exactly one

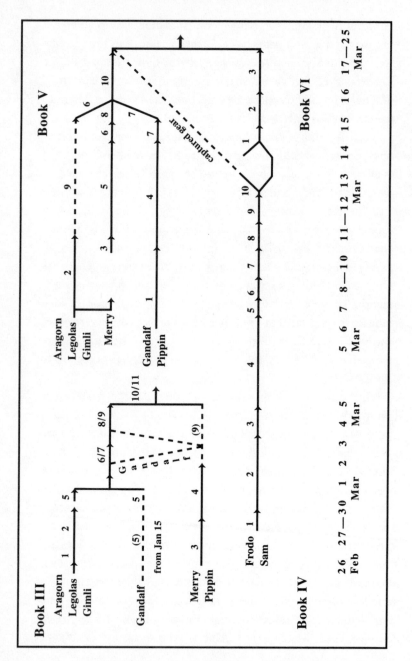

Shire-month, the eight surviving members of the Fellowship (Boromir dies at the very start of *The Two Towers*) are separated. Their adventures are never told for long in strict chronological order, and continually 'leapfrog' each other.

Thus, in the first two chapters of *The Two Towers* we follow Aragorn, Legolas and Gimli from February 26th to February 30th. In chapters 3 and 4 we follow Pippin and Merry from their capture by the Uruk-hai to their meeting with Treebeard; but though these chapters start at (almost) the same time as the first two, the story here goes on further, to March 2nd. Chapters 5 through 8 return us to Aragorn and his companions, soon including Gandalf, picking up at March 1st and continuing this time on to March 5th. Chapter 5 includes Gandalf's necessary 'flashback', explaining his return from the dead, which runs from January 15th. The two groups meet eventually at Isengard, when Gandalf, Aragorn, Théoden and the others find 'two small figures' lying by the shattered gates, one of them asleep and the other peacefully blowing smoke-rings. They are Merry and Pippin, but their appearance is a complete surprise to all including the reader – all, that is, apart from Gandalf, who had met Pippin during his brief detour to Treebeard in chapter 7. The hobbits' explanation of how they got there (like so much of the narration in 'The Council of Elrond') is given in their own narrative, which starts where chapter 4 left off on March 2nd and takes them up to the moment, March 5th. The six members of the Fellowship stay together for two chapters, 10 and 11, but then separate again. The story as far as they are concerned does not resume till the start of *The Return of the King* more than a hundred pages later, when we begin with Gandalf and Pippin (chapters 1 and 4, March 9th to 15th); cut back in between to Aragorn, Legolas and Gimli, who separate from Merry in chapter 2, on March 8th; after which we follow Merry alone, with the Riders, in chapters 3 and 5. All three groups are reassembled at or after 'The Battle of the Pelennor Fields' on March 15th, but once again no one, the reader included, has any idea how Aragorn in particular arrives there. This too has to be told in flashback, the narrative of Legolas and Gimli to Merry and Pippin in V/9 corresponding exactly to the hobbits' narrative to them in III/9. Towards the end of Book V, as towards the end of Book III, the Fellowship is reassembled and moves on together, leaving only Merry behind, to the end of the Book on March 25th.

Meanwhile, of course, and interlaced with all these other threads, there is the narrative of Frodo and Sam in Books IV and VI. This is relatively straightforward, but a point one should note is that Book IV does not end, with the capture of Frodo, at the same time as the corresponding Book III. The first half of *The Two Towers* closes on March 5th, with Pippin and Gandalf riding for Gondor. The second half closes on March 13th, a date which is not reached in the 'alternate' narrative till V/4, by which time Pippin has been in Gondor for some days. As a general rule one may say that none of the five or six major strands of narrative in the central section of *The Lord of the Rings* ever matches neatly with any of the others in chronology: some are always being advanced, some retarded.

Two major effects of this, naturally, are surprise and suspense. It is a surprise to find Pippin and Merry sleeping and smoking in the ruins when they were last seen marching with the Ents on Isengard; it is another to have the Ents determine the Battle of Helm's Deep (something explained only by the hobbits' later flashback); it is a third when the black sails of the Corsairs of Umbar display the White Tree, Seven Stars and crown of Aragorn (to be explained only by Legolas and Gimli's later flashback); obviously it is a major surprise, first to Aragorn, Legolas and Gimli, then to Pippin (and presumably Merry), then to Sam (and presumably Frodo) to have Gandalf return to them from the dead. Perhaps the greatest of the surprises in *The Lord of the Rings*, though, comes at the end of 'The Siege of Gondor' (V/4), when the chief Ringwraith's boast is answered first by the cock-crow and then by 'Great horns of the North wildly blowing. Rohan had come at last.' When last seen the Rohirrim were just leaving the Mark (end of chapter 3), and at the end of chapter 4 we have to go back and follow the Riders through chapter 5 to explain how they arrived. Typically, the end of chapter 5 does not quite match that of chapter 4; it goes on from the hornblast heard by Gandalf and the Ringwraith to Théoden's charge, followed by Éomer, 'the white horsetail on his helm floating in his speed'; and the story does not turn back to Gandalf till chapter 7. And meanwhile, though less easy to pick out, there is a continuous sense of suspense as the reader is left in the dark, for long periods, about the more vital but less conspicuous quest of Frodo and Sam. Not much has been said by critics about the structure of *The Lord of the Rings*, but it is considerably more complex and at least as carefully-integrated as the multiple narrative of Joseph Conrad, for instance, in *Nostromo*. One might feel that a more experienced writer, one who wrote novels or fantasies professionally rather than passionately, would have known not to risk such finesses or trust so much to the ingenuity of his readers: but Tolkien knew no better than to try it.

The main effect of his interlacing technique, however, does not lie in surprise and suspense. What it does is to create a profound sense of reality, of that being *the way things are*. There *is* a pattern in Tolkien's story, but his characters can never see it (naturally, because they are in it). To them the whole story seems chaotic, haunted by bad luck; they are lost in a wilderness metaphorically as well as cartographically, indeed in a 'bewilderment', sometimes in the dark, sometimes in an enchanted wood, frequently guessing wrong as to the meaning of what happens even to them. Aragorn is the first to express this feeling, with his repeated remarks that 'my choices have gone amiss'. As he pursues the Uruk-hai he comes on several puzzles – the dead orcs, Pippin's brooch, Éomer's assurance that there were no hobbits among the dead – which remain unexplained and inexplicable. Even a careful reader may not realize who is 'the old bent man, leaning on a staff' whom Aragorn and his companions see on the edge of Fangorn. It looks very like Gandalf, but when they finally meet Gandalf he tells them it must have been Saruman (though it may have been a 'wraith' of Saruman, possibly projected by Gandalf, see *The Treason of Isengard* p. 428). The fact that Shadowfax returned at the same time to lure away their horses was just a coincidence – if there is such a thing as a coincidence, which Gandalf takes leave to doubt: he remarks at this point how strange it is that their enemies have managed only to 'bring Merry and Pippin with marvellous speed, and in the nick of time, to Fangorn' (and so to stir up the Ents), 'where otherwise they would never have come at all', a remark which bears on Gandalf's theories about 'chance', to be discussed in the next chapter. There were other meetings in Fangorn, however, for Gandalf and Treebeard saw each other, though they did not speak. It was this which made Treebeard reply evasively to the hobbits' tale of Gandalf's death; but the hobbits do not understand this, and in fact Legolas and Gimli (who do know about it) do not enlighten them. The reader has to put three widely-separated conversations together (which perhaps few ever do) to grasp this fact.

On other occasions characters do notice and seize on a connection. In III/5 Legolas remembers that he saw an eagle at the start of III/2, and Gandalf's narrative explains what the eagle was doing. On the other hand, at the end of IV/2 Gollum is convinced that he knows why the Nazgûl have passed over them three times, 'They feel us here, they feel the Precious.' But he is wrong, actually. The third Nazgûl, 'rushing with terrible speed into the West,' is heading for Orthanc to get news of Saruman, alerted perhaps by the burning of the orcs which 'rose high to heaven and was seen by many watchful eyes'. It is the same Nazgûl which passes over the camp at Dol Baran and makes Pippin ask (also wrongly), 'it was not coming for me, was it?', only to be corrected on time and distance by Gandalf. A further cross-connection, which has misled one critic at least, is the scene at the very end of *The Fellowship of the Ring* where Frodo, sitting on the summit of Amon Hen and wearing the Ring, finds himself being shouted at mentally, *Take it off! Take it off! Fool, take it off! Take off the Ring!* Many readers perhaps take this unexplained voice to be some sort of projection from Frodo's own mind, as the last two voices have presumably been (the question is discussed further on pp. 137–8 below); but in fact the third voice is Gandalf's, as one might guess from its asperity: though of course, as far as Frodo and the reader know, Gandalf at that time is dead. It is only later, in III/5 that he says, and then obscurely, 'I sat in a high place, and I strove with the Dark Tower; and the Shadow passed'; nor do the people to whom he is talking understand him.

Other cross-threads in the story include Boromir's horn, split at the end of *The Fellowship of the Ring*, found drifting by Faramir many chapters later, and used by him as a check on Frodo's narrative; Hirgon the errand-rider, who appears fleetingly in three chapters (V/1, V/3, V,5); Faramir's own narrative to his father, 'this is not the first halfling I have seen walking out of northern legends into the Southlands'; and Frodo's mithril coat, displayed by the Mouth of Sauron in V/10 before we have any explanation of how it got there. Perhaps the most significant and the most repeated cross-bearing, however, is provided by the *palantír* of Orthanc. In V/2 Aragorn tells his friends, 'I have looked in the Stone of Orthanc'. They are horrified, and he accepts that it may be that 'I have done ill' in revealing his existence. But in fact he has done well, in the first place prompting Sauron to strike before he is fully ready, as Gandalf realizes some thirty pages later – though his guess is not confirmed till he and Aragorn meet and talk a further sixty pages later, in 'The Last Debate' (V/9). However the major result of Aragorn's decision is never noticed by anyone at all. At a critical moment in Sam and Frodo's travel:

The Dark Power was deep in thought, and the Eye turned inward, pondering tidings of doubt and danger: a bright sword, and a stern and kingly face it saw, and for a while it gave little thought to other things; and all its great stronghold, gate on gate, and tower on tower, was wrapped in a brooding gloom.

The dates of all these events are carefully cross-checked. Frodo and Sam are ignored, or overlooked, on March 16th, after Sauron has had news of the failure at Pelennor Fields and the death of the chief Ringwraith; Aragorn looked in the *palantír*, as he says 'ten days since the Ring-bearer went east from Rauros', i.e. on March 6th; Gandalf guesses that he had done so from Sauron's reaction, 'some five days ago now he would discover that we had thrown down Saruman, and had taken the Stone', i.e. on March 10th. But while the reader can work this out by fitting together the many narratives offered – and also by cross-checking with Tolkien's own detailed chronology in Appendix B – it has to be repeated that the characters do not know, have to guess, sometimes cannot explain, sometimes guess wrong. It is a surprise, for instance, to realize that even at the end of the Ring-quest, when the Ring has been destroyed, Sam and Frodo still do not know that Gandalf has returned from the dead, but think that 'Things all went wrong when he went down in Moria'.

In all this there is a constant irony, created by the frequent gaps between what the characters realize and what the reader realizes – though the reader is of course almost as often in the dark as the characters. But there is also as it were an anti-irony, as one slowly realizes that the characters' frustration, gloom, even approach to despair is at once natural and justified, and also needless and falsified. Things do go wrong, but they could go worse. Even at the worst, there is a vein of proverbial sense which says that one can never be sure: 'Oft evil will shall evil mar', says Gandalf, 'The hasty stroke goes oft astray', says Aragorn, 'a traitor may betray himself, and do good that he does not intend', Gandalf again. Gandalf in particular sometimes draws threads together, commenting for instance (V/8) that it was an important decision by Elrond to allow the junior hobbits to come – they have between them saved Faramir and Éowyn. On the other hand Denethor's despair costs the life of Théoden, but Merry only half-notices ('Where is Gandalf? . . . Could he not have saved the king?', V/6), and though Gandalf is sure that 'if I do [save Faramir], others will die' (V/7), he does not know who they will be. As Gandalf often says in various ways, and he is perhaps thinking of threads and patterns as he does so, 'even the very wise cannot see all ends'.

It may not be possible to draw any certain *correct* conclusion from the confusions and bewilderments of Middle-earth, but it is possible to see one always marked as unequivocally and permanently *wrong*: which is, that there is no point in trying any further. Tolkien makes this declaration through his interlacements, at the same time as he dramatizes the temptation to abandon hope by his separation of narrative threads. But having said that, it is time to leave the map of Middle-earth, and the structure of *The Lord of the Rings*, and consider instead the work's argument, and even (though it is a word Tolkien would not have liked) its ideology.

CHAPTER III

‑‑➤◉◖‑

THE LORD OF THE RINGS (2): CONCEPTS OF EVIL

The concept of the Ring

Most of what has been said in the last two chapters has stressed Tolkien's background in ancient literature. From it one could argue that *The Lord of the Rings* is essentially an 'antiquarian' work, a word now usually used rather patronizingly. The patronage is false, if the antiquarianism is true, and the latter quality does explain a great deal about the charm of Middle-earth. Nevertheless it does not explain why so many readers have found *The Lord of the Rings* so deeply influential, so readily applicable to their own circumstances. Tolkien's work is not just an antiquarian fantasy. If it is still being read (like *Beowulf*) a thousand years after its creation, no perceptive person even in the far future could take it for anything other than a work, a highly characteristic work, of the twentieth century.

One can see this by considering what we are told about the Ring. Tolkien had to do a good deal of work here in modifying what he had said about the ring, Bilbo's ring, the ring not yet imagined as the One Ruling Ring, in the first edition of *The Hobbit*. It comes as rather a shock to anyone who has gathered the story from *The Lord of the Rings* and a later edition of *The*

Hobbit, to go back and read the account of Bilbo's contest with Gollum in the first 1937 edition of *The Hobbit*. The surprising thing there is that Gollum is not all that attached to his 'precious'. He wagers it against Bilbo's life; he loses the riddle-contest; but then he does his best to play fair. When he cannot find the ring (for it is already in Bilbo's pocket), he apologizes profusely for not being able to pay up, and Bilbo, being in a tight corner, accepts Gollum's offer to show him the way out instead. They part on something close to good terms, with Gollum's last words being:

> 'Here'ss the passage . . . It musst squeeze in and sneak down. We dursn't go with it, my preciouss, no we dursn't, gollum.'

From the second edition of 1951 onwards, by contrast, his last words are:

> 'Thief, thief, thief! Baggins! We hates it, we hates it, we hates it for ever!'

Tolkien retained the original and alternative version as the story which Bilbo had told Gandalf and the others, a story in which his claim to the ring was significantly stronger: the fact that Bilbo lied about this is, in *The Lord of the Rings*, an ominous sign that the Ring is gaining power over him, becoming (he uses the same word as Gollum and Isildur), his 'precious'. But this original version of the story contradicts one of the basic facts which we are later told about the Ring, which is that its owners from Isildur on, Gollum included, do not abandon it – it abandons them.

At the heart of *The Lord of the Rings* are the assertions which Gandalf makes in Book I/2, his long conversation with Frodo. If they are not accepted, then the whole point of the story collapses. And these assertions are in essence three. First, Gandalf says that the Ring is immensely powerful, in the right or the wrong hands.

If Sauron regains it, then he will be invincible at least for the foreseeable future: 'If he recovers it, then he will command [all the other Rings of Power] again, even the Three [held by the elves], and all that has been wrought with them will be laid bare, and he will be stronger than ever.' Second, though, Gandalf insists that the Ring is deadly dangerous to all its possessors: it will take them over, 'devour' them, 'possess' them. The process may be long or short, depending on how 'strong or well-meaning' the possessor may be, but 'neither strength nor good purpose will last – sooner or later the dark power will devour him'. Further-more this will not be just a physical take-over. The Ring turns everything to evil, including its wearers. There is no one who can be trusted to use it, even in the right hands, for good purposes: there are no right hands, and all good purposes will turn bad if reached through the Ring. Elrond repeats this assertion later on, 'I will not take the Ring', as does Galadriel, 'I will diminish, and go into the West, and remain Galadriel'. But finally, and this third point is one which Gandalf has to re-emphasize strongly and against opposition in 'The Council of Elrond', the Ring can-not simply be left unused, put aside, thrown away: it has to be destroyed, and the only place where it can be destroyed is the place of its fabrication, Orodruin, the Cracks of Doom.

These assertions determine the story. It becomes, as has often been noted, not a quest but an anti-quest, whose goal is not to find or regain something but to reject and destroy something. Nor can there be any half-measures, attractive as these may seem. Gandalf will not take it, Galadriel will not take it, it would be disastrous to take it to Gondor, as Boromir and Denethor would prefer. One might point out that while all this is perfectly logical, granted the initial assumptions, Gandalf's basic postulates might take a bit of swallowing. Why should we believe them? However, while critics have found fault with almost everything about *The Lord of the Rings*, on one pretext or another, no one to my knowledge has ever quibbled with what Gandalf says about the

Ring. It is far too plausible, and too recognizable. It would not have been so before the many bitter experiences of the twentieth century.

If one fits together the three points which Gandalf makes in this early chapter, it would be a dull mind, nowadays, which did not reflect, 'All power corrupts, and absolute power corrupts absolutely'. This adage was first stated, not in exactly the form just given, in 1887, by Lord Acton. What Lord Acton actually wrote was, 'Power tends to corrupt, and absolute power corrupts absolutely. Great men are almost always bad men . . .' I do not think that many people would have agreed with him much before 1887. The medieval world had its saints' lives, in which the saints used their immense and indeed miraculous power entirely for good purposes; while there is no shortage of evil kings in medieval story, there is rarely any sign that they became evil by becoming kings (though there are some hints to that effect in *Beowulf*). On the whole people probably thought that evil possessors of power were evil by nature, and from the beginning. The nearest thing to Lord Acton's statement in Old English is the proverb, *Man deþ swa he byþ ponne he mot swa he wile*, 'A man does as he is when he can do what he wants', and what this means is that power *reveals* character, not that it alters it. Why have opinions changed?

There is no difficulty in answering this, and the answer proves with particular clarity that Tolkien was not quite as isolated a writer as he is sometimes made out to be. Six years before *The Lord of the Rings* started to be published, George Orwell had published his fable *Animal Farm*, which ends, as everyone knows, with the animals' revolution failing completely, because the pigs had become farmers: 'The creatures outside looked from pig to man, and from man to pig, and from pig to man again; but already it was impossible to say which was which.' The exact applicability of this fable has been furiously disputed (no one wants it to apply to them), but in a way that only makes the point stronger, and it is Gandalf's point: it applies to everybody. All seizures of power, no matter how 'strong or well-meaning' the seizers, will go the same way. That's what power does. Meanwhile, at exactly the same time as the publication of *The Lord of the Rings* William Golding was bringing out his fables, *Lord of the Flies* (1954), and *The Inheritors* (1955), the meaning of which Golding conveniently summarized for commentators in a later essay, 'Fable', in his collection *The Hot Gates*:

> I must say that anyone who passed through those years [of World War II] without understanding that man produces evil as a bee produces honey, must have been blind or wrong in the head.
>
> (*Hot Gates*, p. 87)

So the English choirboys, marooned on an idyllic desert island, invent murder and human sacrifice and create the 'lord of the flies' himself, Beelzebub; in *The Inheritors* our ancestors, Cro-Magnon men, exterminate the gentle and friendly Neanderthals and create an entirely false legend of ogres and cannibals to justify their actions. A very similar if more complex argument was put forward, one might add, by the other great fantasy of the 1950s, T.H. White's *The Once and Future King*, a work which began like Tolkien's with a children's book, *The Sword in the Stone* (1937), but took even longer than Tolkien's to reach termination, appearing as a whole (though still unfinished) in 1958. White's points are too many and too self-doubting to summarize readily, but there is at least no doubt that White saw in humanity a basic urge to destruction, expressed in a work written like *The Lord of the Rings, nationibus in diro bello certantibus*, 'while the nations were striving in fearful war'. Orwell, Golding, White (and several other post-war authors of fantasy and fable): the thought that they expressed in their highly different ways was that people could never be trusted, least of all if they expressed a wish for the betterment of humanity. The major disillusionment of the twentieth century has been over political good intentions, which have led only to gulags and killing fields. That is why what Gandalf says has rung true to virtually everyone who reads it – though it is, I repeat, yet one more anachronism in Middle-earth, and the greatest of them, an entirely modern conviction.

But does Tolkien play fair with this, the very basis of his story? Critics have argued that he does not – pre-eminently Colin Manlove, whose determined attack on Tolkien forms chapter 5 of his 1975 book, *Modern Fantasy*. One should begin by granting that there are indeed several characters who show one stage or another of the creeping corruption which Gandalf fears. We have Bilbo, in the scene in the first chapter when he gets angry with Gandalf as the wizard tries to persuade him to part with (not to hand over) the Ring; and again in 'Many Meetings', when he asks Frodo to let him 'peep at it again', and is transformed for a moment in Frodo's eyes to 'a little wrinkled creature with a hungry face and bony groping hands'. We have Isildur, whose letter discovered in the Gondor archives by Gandalf declares ominously, 'It is precious to me, though I buy it with great pain'. There is Gollum, of course, and throughout. And there is also Boromir, who is doubtful about the wisdom of destroying the Ring in Orodruin from the start, and who in the end breaks the Fellowship because he is convinced that 'True-hearted Men, they will not be corrupted'. Boromir's speech at this point, near the start of the last chapter of *The Fellowship of the Ring*, sends out all the signals which the twentieth century has learned to distrust, from the fascination with power, even if it is 'the power of the Enemy', to the exaltation of 'The fearless', and then immediately 'the ruthless', as the means to victory; finally the self-dramatization of himself as the Leader with 'power of Command', and the naked appeal to force, 'For I am too strong for you, halfling'. Even Sam has a fleeting vision of the same kind, of himself as 'Samwise the Strong, Hero of the Age', when he briefly carries the Ring in VI/1, and as with Bilbo Frodo sees him for a moment, when he is slow to hand over the Ring, as 'a foul little creature with greedy eyes and slobbering mouth'. The danger of carrying the Ring is repeated and insisted on throughout the story in very consistent terms, all bearing out Gandalf's initial assertions.

But, one could say, Tolkien allows exceptions to his own rule. Both Sam and Bilbo do after all hand over the Ring with only momentary reluctance. Other characters show no interest at all in having it or taking it, Merry and Pippin, Aragorn, Legolas and Gimli. Galadriel is aware of it in Lothlórien, and admits that 'my heart has greatly desired' what Frodo offers to give her. She too declares a fantasy like Boromir's and Sam's, 'In place of the Dark Lord you will set up a Queen . . . Dreadful as the Storm and the Lightning! Stronger than the foundations of the earth. All shall love me and despair!' But then she puts aside the temptation with no more than a laugh. In very much the same way Faramir, with Frodo and Sam in his power, and aware of what they are carrying, seems for a moment threatening, but then also laughs and shrugs the temptation off. Finally one has to consider the case of Frodo. Gandalf had told him near the beginning that he could not make him relinquish the Ring, 'except by force, and that would break your mind'. However, as Manlove forcefully points out, in the end, in the scene in the Sammath Naur, Gollum *does* make Frodo give up the Ring, by force, indeed by biting off his finger: but this has no effect on Frodo's mind at all. These apparent contradictions have made critics hostile to the whole fable argue that Tolkien's entire presentation of the origins of evil is flawed: the Ring has bad effects on some people, but no effect at all on others. The plot is being manipulated, not developing logically.

Actually, the doubt expressed in this way can be cleared up by one word, though it is not one that Tolkien uses, and was not recorded by the *OED* till its 1989 edition (the first citation found for it comes only from 1939). The word is 'addictive'. Gandalf's whole argument could be summed up as saying that use of the Ring is addictive. One use need not be disastrous on its own, but each use tends to strengthen the urge for another. The addiction can be shaken off in early stages (which explains Bilbo and Sam), but once it has taken hold, it cannot be broken by will-power alone. On the other hand, if the addiction has not been contracted in the first place (and this explains Galadriel and Faramir, as well as all the other members of the Fellowship), then it has no more power than any other temptation. Moreover, addicts can of course be restrained from their addiction by simple outside force, whether this consists of Gollum's teeth or a locked cell and 'cold turkey' treatment. What Gandalf meant by saying he could not 'make' Frodo hand over the Ring 'except by force, and that would break your mind', was that he could not make him *want* to hand over the Ring except by some unknown mental force, perhaps a kind of hypnosis. But none of this contradicts or detracts from the basic point about the Ring, which is that the very urge to use it is what is destructive: Elrond, or Gandalf, or Galadriel, or Denethor, if they owned it, would begin with the best of intentions, but would come to enjoy having their intentions achieved, the use of power itself, and would end as dictators over others, enslaved to themselves, unable to give up or go back.

Wraiths and shadows: Tolkien's images of evil

There is something extremely convincing, for very many people, in Tolkien's presentation of evil; but it is worth re-stressing that his concern with the topic is highly contemporary, and by no means unique. Many authors of the mid-twentieth century were obsessed with the subject of evil, and produced unique and original images of it. I have mentioned already Orwell's torturer O'Brien, in *Nineteen Eighty-Four*, declaring, 'If you want a picture of the future, imagine a boot stamping on a human face – for ever'; and Ursula Le Guin's parable of 'The Ones Who Walk Away From Omelas', with its shining city whose power and beauty depend entirely on the continuous and conscious tormenting of an idiot child; to which one can add Kurt Vonnegut's Billy Pilgrim, working in the 'corpse mines' of Dresden, with their stink 'like roses and mustard gas'; or T.H. White's Merlyn denouncing humanity as:

> '*Homo ferox*, the Inventor of Cruelty to Animals, who will
> ... burn living rats, as I have seen done in Eriu, in order
> that their shrieks may intimidate the local rodents; who will
> forcibly degenerate the livers of domestic geese, in order to
> make himself a tasty food; who will saw the growing horns
> of cattle, for convenience in transport; who will blind gold-
> finches with a needle, to make them sing; who will boil
> lobsters and shrimps alive, although he hears their piping
> screams; who will turn on his own species in war, and kill
> nineteen million every hundred years' (etc.)
>
> (*The Book of Merlyn*, section 5)

All these images are based, sometimes obviously as with Vonnegut, sometimes less obviously as with Le Guin, on personal or recent experience. The authors are trying to explain something at once deeply felt and rationally inexplicable, something furthermore felt to be entirely novel and not adequately answered by the moralities of earlier ages (keen medievalists though several of these authors were). The end of the quotation above from White suggests that this 'something' is connected with the distinctively twentieth-century experience of industrial war and impersonal, industrialized massacre; and it is probably no coincidence that most of the authors concerned (Tolkien, Orwell, Vonnegut, but also Golding, and Tolkien's close colleague C.S. Lewis) were combat veterans of one war or another. The life experiences of many men and women in the twentieth century have left them with an unshakable conviction of something wrong, something irreducibly evil in the nature of humanity, but without any very satisfactory explanation for it. Nor can they find such an explanation in the literature of previous eras: Billy Pilgrim's friend Rosewater in Vonnegut's *Slaughterhouse-Five* agrees that, 'everything there was to know about life was in *The Brothers Karamazov*, by Feodor Dostoyevsky. 'But that isn't *enough* any more'. Twentieth-century fantasy can be seen as above all a response to this gap, this inadequacy. One has to ask in what ways Tolkien's images are original, individual, and in what ways typical, recognizable.

The orcs, whom we meet or overhear several times in *The Lord of the Rings*, form one image, and there is a conclusion to be drawn from them (see the next section). However, they are relatively low-ranking evil-doers, what Tolkien called in his *Beowulf* lecture 'the infantry of the old war'; and in some ways they resemble fairly conventional fairy-tale images, like the 'goblins', which was Tolkien's original word for them. More individual and more original is Tolkien's concept of the 'Ringwraith'. This is, one has to say, a word of exactly the same type as 'wood-wose' or **hol-bytla*: a compound, the first element completely familiar, the second more mysterious. What is a 'wraith'? If one looks the word up in the *OED* one finds a puzzle of just the kind which always attracted Tolkien's attention. The dictionary has no suggestion about the word's etymology, but comments 'Of obscure origin'. As for its meaning, the *OED* gives two senses, which appear to contradict each other, and cites the same text, Gavin Douglas's 1513 translation of Virgil's *Aeneid* into Scots, as the source for both. I have no doubt that Tolkien and the other Inklings – for Lewis has a very clear image of a fictional wraith as well – discussed the matter, and in the end found a solution which makes sense both of Douglas's old text, and of the modern reality to which 'wraiths' refer. (For a comment on Lewis, see my article in the Clark and Timmons collection in the 'List of References' below).

To begin with the etymology of 'wraith', an obvious suggestion which the *OED* compilers should have thought of is that it is a form derived from the Old English verb *wríðan*, 'writhe'. This is a class 1 strong verb, exactly parallel to *rídan*, 'ride', and if it had been common enough to survive in full form, we would still say 'writhe – wrothe – writhen', as we do 'ride – rode – ridden' or 'write – wrote – written' (Tolkien does in fact use the form 'writhen', see Blackwelder's *Tolkien Thesaurus*). It is characteristic of verbs like 'ride' or 'write' to form other words by vowel-change, like 'road' from 'ride' or 'writ' from 'write'. 'Writhe' has given rise to several: 'wreath' (something that is twisted), but less obviously and more suggestively, 'wroth' (the old adjective meaning 'angry'), and 'wrath' (the corresponding noun which still survives). What has anger got to do with writhing, with being twisted? Clearly – and there are other parallels to this – the word is an old dead metaphor which suggests that wrath is a state of being twisted up inside (an Inkling thesis expressed by Owen Barfield and mentioned by Tolkien, see *Letters* p. 22. The word *wraithas*, 'bent', was also of special importance to Tolkien's personal myth of 'the Lost Road', see pp. 287–8 below.)

That Tolkien was aware of this sort of variation between the physical and the abstract is suggested by a word Legolas uses in 'The Ring Goes South'. There, when the Fellowship's attempted crossing of Caradhras is foiled by the snow, Legolas goes ahead to scout out their retreat. He returns to say that the snow does not reach far, though he has not brought the sun back with him: 'She is walking in the blue fields of the South, and a little *wreath* of snow on this Redhorn hillock troubles her not at all'. By 'wreath' here Legolas clearly means something like 'wisp', something barely substantial, and though the *OED* does not record it, that is also part of the meaning of 'wraith' – one could say, 'a wraith of mist', 'a wraith of smoke', just as Legolas says 'a wreath of snow'. It seems likely, then, that 'wraith' is a Scottish form derived from *wríðan* in exactly the same way as 'raid' is derived from *rídan*.

Meanwhile the two Gavin Douglas quotations from which the *OED* derives its two senses are these. To illustrate sense 1, 'An apparition or spectre of a dead person: a phantom or ghost', the *OED* offers Douglas, 'In diuers placis The wraithis walkis of goistis that are deyd'. For sense 1b, though, 'An immaterial or spectral appearance of a living being', it offers Douglas again, 'Thidder went this wrath or schaddo of Ene' (i.e. Virgil's hero Aeneas). The obvious question is, are wraiths, then, alive or dead, for Douglas uses the word both ways? And, one might add, are they material or immaterial? The latter is suggested by the equation with 'shadow' (another important word for Tolkien), and by the idea that wraiths and wreaths are defined by their shape more than by their substance, a twist, a coil, a ring; the former, however, by the fact that wraiths can be wraiths *of* something, even if that something is as fluid (but not insubstantial) as snow or mist or smoke. Tolkien's Ringwraiths, of course, answer all the questions posed, and also demonstrate once more that apparent mistakes or contradictions in old poems may simply indicate an understanding that the self-confident nineteenth- and twentieth-century dictionary compilers had not reached. Are the Ringwraiths alive or dead? Gandalf says early on that they were once men who were given rings by Sauron, and so 'ensnared . . . Long ago they fell under the dominion of the one [Ring], and they became Ringwraiths, shadows under his great Shadow, his most terrible servants'. Much later, in 'The Battle of the Pelennor Fields', we learn that the Lord of the Nazgûl, the chief Ringwraith, was once the sorcerer-king of Angmar, a realm overthrown more than a thousand years in the past. He *ought*, then, to be dead, but is clearly alive in some way or other, and so positioned neatly between the two meanings given by the *OED*. As for being material or immaterial, he is in a way insubstantial, for when he throws back his hood, there is nothing there. Yet there must be *something* there, for 'he had a kingly crown; and yet upon no head visible was it set'. He and his fellows can furthermore act physically, carrying steel swords, riding horses or winged reptiles, the Lord of the Nazgûl wielding a mace. But they cannot be harmed physically, by flood or weapon – except by the blade of Westernesse taken from the barrow-wight's mound, wound round with spells for the defeat of Angmar. It is the spells that cleave 'the undead flesh', not the blade itself. So the Ringwraiths are just like mist or smoke, both physical, even dangerous and choking, but at the same time effectively intangible.

All this is highly original. But the important question is, how far is it recognizable, even psychologically plausible? And here the answer returns us very firmly to the twentieth century. Tolkien did not perhaps develop his image of the Ringwraiths very quickly, for as has been said above, the Black Riders to begin with make relatively little impact. In 'The Council of Elrond', though, Boromir gives them what is to be one of their leading characteristics, the ability to create panic: wherever the 'great black horseman' came, 'a madness filled our foes, but fear fell on our bravest'. This is increasingly what the wraiths do from the time the Fellowship emerges from Lothlórien. When they pass overhead, over Sam and Frodo, over the Riders, over Gondor, we have some combination of the same elements: shadow, cry, freezing of the blood, fear. The moment when Pippin and Beregond hear the Black Riders and see them swoop on Faramir in 'The Siege of Gondor', V/4, is typical:

> Suddenly as they talked they were stricken dumb, frozen as it were to listening stones. Pippin cowered down with his hands pressed to his ears; but Beregond . . . remained there, stiffened, staring out with starting eyes. Pippin knew the shuddering cry that he had heard: it was the same that he had heard long ago in the Marish of the Shire, but now it

was grown in power and hatred, piercing the heart with a poisonous despair.

The last phrase is a critical one. The Ringwraiths work for the most part not physically but psychologically, paralysing the will, disarming all resistance. This may have something to do with the process of becoming a wraith yourself. That can happen as a result of a force from outside. As Gandalf points out, explaining the Morgul-knife, if the splinter had not been cut out, 'you would have become a wraith under the dominion of the Dark Lord'. But more usually the suspicion is that people *make themselves into wraiths*. They accept the gifts of Sauron, quite likely with the intention of using them for some purpose which they identify as good. But then they start to cut corners, to eliminate opponents, to believe in some 'cause' which justifies everything they do. In the end the 'cause', or the habits they have acquired while working for the 'cause', destroys any moral sense and even any remaining humanity. The spectacle of the person 'eaten up inside' by devotion to some abstraction has been so familiar throughout the twentieth century as to make the idea of the wraith, and the wraithing-process, horribly recognizable, in a way non-fantastic.

The realism of this image of evil is increased by the examples we have of people on their way to becoming wraiths themselves. We have just the start of this, enough to be ominous, in the cases of Bilbo and Frodo, and the others mentioned above. Gollum is much further along the road, though in *The Lord of the Rings* Gollum, detached from the Ring many years before, is possibly beginning to recover, as is shown by the fact that he has started to call himself by his old name, Sméagol, the name he had when he used to be a hobbit, and is also occasionally and significantly able to say 'I'. There is a striking dialogue between what one might call his hobbit-personality (Sméagol) and his Ring-personality (Gollum, 'my precious') in 'The Passage of the Marshes', which makes the point that the two are at least connected: one can imagine the one developing out of the other, pure evil growing out of mere ordinary human weakness and selfishness.

However, the best example of 'wraithing' in *The Lord of the Rings* must be Saruman. As was pointed out earlier, his language and behaviour are the most contemporary of any in 'The Council of Elrond', or indeed in the whole work. Saruman's goals are knowledge (no one can object to that); organization in the service of knowledge (there are certainly many researchers, and far more administrators, who see this as desirable); but finally control. In the pursuit of control Saruman is prepared to co-operate with forces he knows perfectly well are evil, but which he thinks he can use for his own much more admirable purposes, and later suppress or discard. The failure of beliefs like this is all too familiar from war after war, and alliance after alliance, during the past century. Moreover Saruman's main advantage, we learn in 'The Voice of Saruman' (III/10) is indeed his voice:

> Those who listened unwarily to that voice could seldom report the words that they heard; and if they did, they wondered, for little power remained in them. Mostly they remembered only that it was a delight to hear the voice speaking, all that it said seemed wise and reasonable, and desire awoke in them by swift agreement to seem wise themselves. When others spoke they seemed harsh and uncouth by contrast . . .

Different people will have different real-life experiences to match this too, but it is again a common experience in the world of the twentieth century to find oneself enmeshed in some professional jargon, whether it is that of Vietnam generals with their body-counts or of literary theorists with their *différances* and *ratures*, and to be unable to break free of it, or to shake off the assumptions it tacitly embodies; the experience goes further back than either of the examples just cited, as one can tell from Orwell's repeated criticisms of early twentieth-century military and political language. Saruman is becoming a wraith, then, partly by merging himself with his own cause, discarding any sense of means in pursuit of some increasingly impossible end, and partly by the self-deceptions of language. He too becomes physically a wraith in the end, for when Wormtongue cuts his throat, the wraith rises from him:

> about the body of Saruman a grey mist gathered, and rising slowly to a great height like smoke from a fire, as a pale shrouded figure it loomed over the Hill. For a moment it wavered, looking to the West; but out of the West came a cold wind, and it bent away, and with a sigh dissolved into nothing.

The body that is left once the 'mist' and the 'smoke' have departed seems in fact to have died many years before, becoming only 'rags of skin upon a hideous skull'. There was still some humanity in Saruman – the figure which wavers, looking towards the West, is perhaps hoping for some forgiveness from the Valar, as the dissolving sigh perhaps indicates some sort of grief or repentance – but it had been steadily eaten up.

By what? C.S. Lewis might have replied, by nothing. One of the striking and convincing assertions made by his imagined devil, Screwtape, is that nowadays the strongest temptations are not to the old human vices of lust and gluttony and wrath, but to new ones of tedium and solitude. At the end of number XII of *The Screwtape Letters* Screwtape remarks that Christians describe God as the One 'without whom Nothing is strong', and they speak truer than they know, he goes on, for:

> Nothing is very strong: strong enough to steal away a man's best years not in sweet sins but in a dreary flickering of the

mind over it knows not what and knows not why, in the
gratification of curiosities so feeble that the man is only
half aware of them ... or in the long, dim labyrinth of
reveries that have not even lust or ambition to give them
a relish.

Sinners of this kind, of course, hate all those who appear to have
'got a life', in the revealing modern idiom; it is essential that they
persuade others to join them in their dreariness and despair. And
so we have the many modern literary images of evildoers as above
all 'hollow' (T.S. Eliot wrote a poem called 'The Hollow Men');
of evil as essentially pointless or bureaucratic (see Vonnegut's
Slaughterhouse-Five, or Heller's *Catch-22*); of the power of language to conceal unmistakable evil (the inhabitants of Le Guin's
Omelas, most of them, talking themselves out of what they have
seen with their own eyes); of something dreadful underneath the
routines of daily life, as in Conrad's prosaic Marlow coming upon
the 'heart of darkness' and Kurtz's never-explained 'The horror,
the horror'. No one ever wrote anything like that in the Middle
Ages. Tolkien may have taken his word 'wraith' from the sixteenth century and Gavin Douglas, but the concept of the Ringwraiths themselves, and the hints as to how you get to be one,
are responses to something found in his own, and our, life-
experience. That is what has given them, not their literary originality, but their dreadful conviction.

Two views of evil

The word which goes with 'wraith' from Gavin Douglas's time
is 'shadow', and it is a word which Tolkien uses repeatedly and
pointedly. In the verse of the rings which Gandalf quotes to Frodo
in 'The Shadow of the Past', the concluding lines are:

One Ring to rule them all, One Ring to find them,
One Ring to bring them all, and in the darkness bind them,
In the land of Mordor where the shadows lie.

When Gandalf falls into the abyss, Aragorn says that he 'fell into
shadow'; Gandalf says that if they lose, 'many lands will pass
under the shadow'; sometimes 'the Shadow' becomes a personifi-
cation of Sauron, as when Frodo tells Sam that 'the Shadow can
only mock, it cannot make . . . not real new things of its own'.
The last statement goes far towards explaining why Tolkien used
the word so often and with such emphasis. One might think that
the main associations of 'shadow' are darkness, or menace, or
perhaps oblivion, but the real point may be a more metaphysical
one. Do shadows exist? In the Old English *Solomon and Saturn*
poem from which Tolkien drew Gollum's riddles, Solomon
indeed asks Saturn, 'What is that is not?' And though the answer
is expressed riddlingly, it contains the word *besceadeð*, 'shadows'
(here a verb). Saturn seems to be saying that shadows both are
and aren't. Aren't, in that a shadow is not a thing, but an absence
caused by a thing. Are, in that they have shapes, and physical
effects, like cold and dark. In folklore at least they can be detached,
even stolen. Particularly ominous, therefore, is the slight variation
on the line from the rings-verse given by Sam, when he recites
the elvish poem about Gil-galad in I/11. This ends (my emphasis):

For into darkness fell his star,
In Mordor where the shadows *are*.

Just as the wraiths are both substantial and insubstantial, in
Mordor (though Sam does not realize the ominousness of what
he says), absence can take on a kind of life, can become presence
– as it does for instance in Milton's presentation of Death in
Paradise Lost II, 666–73, also a 'shape' poised between 'substance'

and 'shadow', and like the chief Ringwraith, bearing 'the likeness of a kingly crown' on 'what seemed its head'.

By saying things like this, however, Tolkien sets up a running ambivalence throughout the whole of *The Lord of the Rings*, which acts as an answer at once orthodox and questioning to the whole problem of the existence and source of evil in a universe created (as both Tolkien and Milton were sure it was) by a benevolent God. One can sum Tolkien's characteristically twentieth-century position up by saying that there are two opinions about the nature of evil, both old, both deep-rooted, both still relevant, neither easy to deny, but apparently irreconcilably in contradiction. One is that of orthodox Christianity, repeated and put into modern language by, for instance, Tolkien's close friend and associate C.S. Lewis, whose exposition of it in *Mere Christianity* was composed at the same time as Tolkien was writing the first chapters of *The Lord of the Rings*, and eventually published in 1952. One of Lewis's avowed motives in writing the book (in which 'mere' means 'common' or 'central') was to state doctrines which both he, an Ulster Protestant, and Tolkien, a Catholic, could agree on. Furthermore, as both Tolkien and Lewis would certainly have known, the most famous statement of this view of evil was made in a work written by a Christian, which however never at any point mentions Christ or any specifically Christian doctrine, trying at all times to reach its conclusions through logic alone: the *De Consolatione Philosophiae* written in the sixth century by Boethius, a Roman senator at the time of writing under sentence of death on charges of plotting to restore Imperial rule (a sentence in the end carried out: Boethius was tortured to death in AD 524 or 525).

The Boethian view is this: there is no such thing as evil. What people identify as evil is only the absence of good. Furthermore people in their ignorance often identify as evil things (like being under sentence of death) which are in fact and in the long run, or in the divine plan, to their advantage. Philosophy tells Boethius that 'all fortune is certainly good', *omnem bonam prorsus esse fortunam*. Corollaries of this belief are, as Frodo says to Sam in 'The Tower of Cirith Ungol', that evil cannot create, 'not real new things of its own', and furthermore it was not created; it arose (and here we switch over to 'Mere Christianity') when human beings exercised their own free will in withdrawing their service and their intentions from God; in the end, and when the divine plan has been fulfilled, all evils may be annulled, cancelled, brought to good, as the Fall of Man was by the Incarnation and Death of Christ. As all readers of Boethius have observed – and his translators into English have included King Alfred, Chaucer, and Queen Elizabeth the First – whatever one may think of the truth of Boethius's opinions, no one can deny his fortitude in writing them on Death Row while waiting for execution. His view of the non-existence of evil has great authority, both in its own right and through its ratification by orthodox Christianity.

There is also a certain amount of evidence for it, put into colloquial language by Lewis and fictionalized by Tolkien through the rather unlikely medium of the orcs. To put Lewis's argument first, a point he made with characteristic simplicity at the start of *Mere Christianity* is that even evil-doers are liable to excuse themselves in terms of what is good: breakers of promises insist that they do so because circumstances have changed, murderers claim that they were provoked, atrocities are excused as retaliation for earlier atrocities, and so on. Lewis claims that 'in reality we have no experience of anyone liking badness just because it is bad'; and since bad and good are not symmetrical in this way, evil is an absence, as Boethius said, and also 'a parasite, not an original thing', rather as Frodo had said. The argument remains, however, rather abstract. One can see Tolkien here and there doing his best not only to make it more realistic, but even, for those with a robust sense of humour, even funny.

A clear but unnoticed example comes from the orcs. We hear orcs talking six times in *The Lord of the Rings*; I consider their conversations in more detail in the article in the Clark and Timmons collection mentioned already, but the point can be made from one conversation alone. In the last chapter of *The Two Towers* Frodo has fallen paralysed by the venom of Shelob the spider, and although Sam takes the Ring from him, he then falls into the hands of the orcs. Sam, wearing the Ring, can hear the dialogue of the two orc-leaders, Gorbag from Minas Morgul and Shagrat from Cirith Ungol. Gorbag warns Shagrat that while they have captured the one 'spy', Frodo, it is clear that someone else, presumably 'a large warrior ... with an elf-sword', wounded Shelob and is still loose. The 'little fellow' they have caught:

> 'may have had nothing to do with the real mischief. The big fellow with the sharp sword doesn't seem to have thought him worth much anyhow – just left him lying: regular elvish trick.'

There is no mistaking the disapproval in Gorbag's voice. He is convinced that it is wrong, and contemptible, to abandon your companions. Furthermore it is characteristic of the other side, a 'regular elvish trick', they do it all the time. Nearly everything Gorbag says is factually wrong, and it is less than a page before this orcish view of morality is also exposed. For Shagrat knows something which Gorbag doesn't, which is that Shelob has 'more than one poison'. She usually paralyses her prey rather than killing it outright. Shagrat asks:

> 'D'you remember old Ufthak? We lost him for days. Then we found him in a corner; hanging up he was, but he was wide awake and glaring. How we laughed! she'd forgotten him, maybe, but we didn't touch him – no good interfering with Her.'

What can one say but 'regular orcish trick'? It is true that it is Gorbag who expresses disapproval of abandoning one's companions, when other people do it, and Shagrat who laughs at doing exactly that, when he does it, but on this matter there seems to be no disagreement between them. Orcs here, and on other occasions, have a clear idea of what is admirable and what is contemptible behaviour, which is exactly the same as ours. They cannot revoke what Lewis calls 'the Moral Law' and create a counter-morality based on evil, any more than they can revoke biology and live on poison. They are moral beings, who talk freely and repeatedly of what is 'good', meaning by that more or less what we do. The puzzle is that this has no effect at all on their actual behaviour, and they seem (as in the conversation quoted) to have no self-awareness or capacity for self-criticism. But these are human qualities too. The orcs, though low down on the scale of evil, the mere 'infantry of the old war', quite clearly and deliberately dramatize what I have called the Boethian view: evil is just an absence, the shadow of the good.

The trouble with this view is that it is both highly counter-intuitive, and in many circumstances extremely dangerous. One might, for instance, conclude that the proper response to it, if you accepted it, would be to become a conscientious objector, and to refuse to resist what appears to be evil on the ground that this is just a misapprehension. Evil after all is, according to Boethius, more harmful to the malefactor than to the victim and those who do it (or appear to do it) are more to be pitied than feared or fought. King Alfred, dictating his Old English translation of Boethius in the intervals of fighting a desperate war against heathen Vikings, in which he hanged both pirates taken prisoner and also on one occasion his own rebellious monks, certainly found it impossible to go along with Boethius all the way; while at the time that Tolkien was writing *The Lord of the Rings*, surrender to his country's enemies would have meant handing over not only himself but many others to the whole apparatus of concentration camps, gas-chambers, and mass murder. A brave man might be prepared to be Boethian himself. But did he have the right to impose the results of that stance on others more defenceless? Neither Tolkien nor King Alfred would have thought so.

In any case there is an alternative tradition in Western thought, which has never risen to the status of being official, but which generates itself spontaneously from common experience. This says that while it may be all very well to make philosophical statements about evil, nevertheless evil *does* exist, and is not merely an absence; and what is more, it has to be resisted and fought, not by all means available, but by all means virtuous; and what is even more, *not* doing so, in the belief that one day Omnipotence will cure all ills, is a dereliction of duty. The danger of this opinion is that it swerves towards being a heresy, Mani-chaeanism, or Dualism: the belief that the world is a battlefield, between the powers of Good and Evil, equal and opposite – so that, one might say, there is no real difference between them, and it is a matter of chance which side one happens to choose.

The Inklings, as it happens, may have had a certain tolerance for Manichaeanism – in *Mere Christianity* II/2 Lewis awards Dual-ism second place, so to speak, after Christianity, before going on to make the case against it – but Tolkien certainly less than Lewis. It annoyed him very much when the reviewer for the *Times Literary Supplement* asserted that in *The Lord of the Rings* all that the good and the bad sides did was try to kill each other, so that they could not be told apart: 'Morally there seems nothing to choose between them' (this comes from a letter in the *TLS* for 9th December 1955, in which the reviewer, Alfred Duggan, defended himself against challenge; Tolkien later corresponded with David Masson, who had made the challenge precisely over the issue of the (dis)similarity of the good and evil sides). Tolkien was a more orthodox Christian than Lewis, and less tolerant of anything like heresy. Nevertheless, his education, his faith, and the circumstances of his time, all set up what seemed to be a deep-seated contradiction between Boethian and Manichaean opinions, between authority and experience, between evil as an absence ('the Shadow') and evil as a force ('the Dark Power'). In *The Lord of the Rings* this contradiction drives much of the plot. It is expressed not only through the paradoxes of wraiths and shadows, but also through the Ring.

Evil and the Ring

The Ring's ambiguity is present almost the first time we see it, in 'The Shadow of the Past', when Gandalf tells Frodo, 'Give me the ring for a moment'. Frodo unfastens it from its chain and, 'handed it slowly to the wizard. It felt suddenly very heavy, as if either it or Frodo himself was in some way reluctant for Gandalf to touch it.'

Either it *or* Frodo. It may not seem very important to know which of these alternative explanations is true, but the difference is the difference between the world-views I have labelled above as 'Boethian' and 'Manichaean'. If Boethius is right, then evil is internal, caused by human sin and weakness and alienation from God; in this case the Ring feels heavy because Frodo (already in the very first stages of addiction, we may say) is unconsciously reluctant to part with it. If there is some truth in the Manichaean view, though, then evil is a force from outside which has in some way been able to make the non-sentient Ring itself evil; so it is indeed the Ring, obeying the will of its master, which does not want to be identified. Both views are furthermore perfectly convincing. In the earlier scene of Bilbo's inability to part with the Ring – not realizing it's in his pocket, getting angry when pressed, unable to make up his mind, dropping the envelope with the Ring on the floor – all readers realize that these are not accidents, but manifestations of Bilbo's own unconscious wishes:

Freudianism has taught us all at least that much. However the whole plot of *The Lord of the Rings* is permeated with the idea of the will of Sauron operating at a distance, stirring up evil forces, literally animating the Ringwraiths and even the orcs; Gandalf talks repeatedly of the Ring as animate, betraying Isildur, abandoning Gollum, and says in explanation that according to Bilbo the Ring 'needed looking after . . . it shrank or expanded in an odd way, and might suddenly slip off a finger where it had been tight'. The ideas that on the one hand the Ring is a sort of psychic amplifier, magnifying the unconscious fears or selfish-nesses of its owners, and on the other that it is a sentient creature with urges and powers of its own, are both present from the beginning, and correspond to the internal/Boethian and external/Manichaean theories of evil.

The ambiguity is more prominent and more important in later scenes. Frodo puts on the Ring six times in *The Lord of the Rings*. The first time is in the house of Tom Bombadil. This does not seem to count, for Tom, characteristically, is quite unaffected: he neither becomes invisible himself when he puts it on nor fails to see Frodo when *he* puts it on. The next time is in the *Prancing Pony*, when Frodo feels a 'desire . . . to slip it on and vanish out of the whole silly situation'. This, of course, could be entirely his own doing; but 'It seemed to him, somehow, as if the suggestion came to him from outside'. In any case 'He resisted the temptation firmly'. He makes a speech, sings a song, and then, falling off the table on which he has been capering, finds he has put on the Ring. By accident? Frodo at least works out an explanation of how this could have happened. But at the same time 'he wondered if the Ring itself had not played him a trick; perhaps it had tried to reveal itself in response to some wish or command that was felt in the room'. We never learn the truth about this, and the second explanation does not seem especially plausible. Who in the room could have given such a command? The likes of Bill Ferny seem too low-rank and too ignorant to be capable of projecting such orders. But this is not the case on Weathertop, when the Ring-wraiths attack.

Here the Manichaean view is much more evident. Frodo remembers all his warnings, but 'something seemed to be compelling him' to disregard them. The situation is different, again, from the moment in the Barrow-wight's mound, when Frodo thought for a moment of using the Ring to escape, but put the thought aside without difficulty. On Weathertop he has 'no hope of escape . . . he simply felt that he must take the Ring and put it on his finger'. He struggles against the urge for a while, but in the end 'resistance became unbearable'. The feeling here is that Frodo's will has just been overpowered by superior force, no doubt that of the wraiths, using some mental power of the sort Gandalf hinted at. And yet, and on the other hand, the word used at the start of the attack (just as in the *Prancing Pony*) is 'temptation': Frodo is tempted. Furthermore, we are told that it would have made a difference if he had yielded to the temptation. Gandalf says later on that his heart was not pierced by the Morgul-knife 'because you resisted to the last'. He might mean just that Frodo dodged, shouted, struck out, in an entirely physical sense putting the Ringwraith off his aim. But more likely there is a psychological sense. The knife works by subduing the will, and if the will does not co-operate it works less well – though it does not lose its power entirely and altogether, as it would if evil were entirely a matter of inner temptations. Gandalf keeps up the ambiguity of the scene by remarking that 'fortune or fate have helped you . . . not to mention courage'. But here he clearly means not either/or but both, fate *and* courage: the same may be true of the nature of the Ring.

Frodo uses the Ring twice on Amon Hen (II/10), and both times he has to, first to escape Boromir, then to get away from the Fellowship without being noticed. On the first occasion, though, he sees the Eye of Sauron, and becomes aware that it is looking for him. And as he does so:

He heard himself crying out *Never, never!* Or was it: *Verily, I come, I come to you?* He could not tell. Then as a flash from some other point of power there came to his mind another thought: *Take it off! Take it off! Fool, take it off! Take off the Ring.*

The two powers strove in him. For a moment, perfectly balanced between their piercing points, he writhed, tormented. Suddenly he was aware of himself again. Frodo, neither the Voice nor the Eye: free to choose and with one remaining instant in which to do so. He took the Ring off his finger.

This is an especially mysterious scene on first reading, though it is cleared up slightly when we learn (as has been said above) that the third voice is Gandalf's, in a 'high place' somewhere striving against the mental force of the 'the Dark Power'. But whose are the other two voices? The first one seems to be 'himself', i.e. Frodo. The second one could be, perhaps, the voice of the Ring: the sentient creature obeying the call of its maker, Sauron, as it has been all along. Or could it be, so to speak, Frodo's subconscious, obeying a kind of death-wish, entirely internal but psychically amplified by the Ring? For that, after all, is how we are told the Ring works. It gets a hold on people through their own impulses, towards pity or justice or knowledge or saving Gondor, and gives them the absolute power that corrupts absolutely. There has to be something there for it to work on; but, like the worms in Bilbo's father's proverb, everyone has some weak spot. They may 'writhe' between the external and internal powers, but that is surely how one gets to be a 'wraith'.

The Manichaean images of the Ring become stronger as it moves closer to Mordor. Sam's uses of it – he puts it on twice – are conditioned by immediate necessity, like Frodo's on Amon Hen, but he too feels it both as an external power, 'untameable save by some mighty will', and as an inner temptation. Here, though, it seems obvious that the temptation to become 'Samwise the Strong, Hero of the Age' is mostly the Ring's, amplifying whatever petty selfish urge it can find. Sam hardly feels the temptation, and puts it aside as a 'shadow', mere 'phantoms'. In a similar way, on the Stairs of Cirith Ungol, Frodo hides from the Lord of the Nazgûl, but is sensed by him. Frodo feels 'the beating upon him of a great power from outside', which takes his hand and moves it 'inch by inch towards the chain upon his neck'. But this time 'There was no longer any answer to that command in his own will', so that he can force his hand back, to the phial of Galadriel. 'No longer' of course implies that there had been such an answer previously, on Amon Hen, on Weathertop, or in the *Prancing Pony*. But this time there is no doubt that the 'power' is from 'outside'.

The last and critical scene, however, is the one on Mount Doom, in the chambers of the Sammath Naur. In the approach to this the sense of an outside power has grown stronger and stronger. Sam sees Frodo's hand creep again and again towards the Ring, only to be withdrawn 'as the will recovered mastery'. It is a surprise, then, that when Frodo at last glimpses the Eye, reaches for the chain and the Ring, and whispers to Sam, 'Hold my hand! I can't stop it', Sam can take his hand away and hold it without effort, indeed 'gently'. The force that is operating on Frodo is not a physical one, like magnetism, which would be unaffected by personality; what is unstoppable to Frodo is imperceptible to Sam. In the same way, the Ring is a crushing burden to Frodo, but when Sam picks him up, expecting to feel the same 'dreadful dragging weight of the accursed Ring', it is no weight at all. Meanwhile the outside power is having an effect on Sam, but it operates once again (as in the scene on Amon Hen) by creating a kind of dialogue. Sam finds himself holding 'a debate with himself'. One voice is optimistic, determined, set on destroying the Ring. The other voice – it is 'his own voice', but it twice calls him 'Sam Gamgee', as if it was someone else – says he can't go on, doesn't know what to do, and 'might just as well lie down now and give it up'. Whose voice is this? It could, of course, just be Sam's own feelings of downheartedness: most people talk to themselves mentally at some point. On the other hand, it could be the Ring, once more amplifying inner feelings and this time giving them a voice. When Sam finally rejects the second voice, whoever's it is, the ground shakes and rumbles, as if some outside power had recognized and resented his decision. All this builds up to the question of what makes Frodo fail at the last hurdle. He reaches the Sammath Naur, leaving Sam behind to deal with Gollum, and when Sam follows him in, he finds that even the phial of Galadriel is no longer any use to him. In this place, 'the heart of the realm of Sauron . . . all other powers were here subdued'. At that moment, standing on the very edge of the Crack of Doom, Frodo gives up. His words are:

> 'I have come . . . But I do not choose now to do what I came to do. I will not do this deed. The Ring is mine.'

With that he puts it on for the sixth and final time. It is a vital question to know whether Frodo does this because he has been made to, or whether he has succumbed to inner temptation. What he says suggests the latter, for he appears to be claiming responsibility very firmly: 'I will not . . . the Ring is mine.' Against that, there has been the increasing sense of reaching a centre of power, where all other powers are 'subdued'. If that is the case, Frodo could no more help himself than if he had been swept away by a river, or buried in a landslide. It is also interesting that Frodo does not say, 'I choose not to do', but 'I do not choose to do'. Maybe (and Tolkien was a professor of language) the choice of words is absolutely accurate. Frodo does not choose; the choice is made for him.

The question becomes an academic one, of course, in that the result is achieved by Gollum, fulfilling Frodo's own words a few moments before, 'If you touch me ever again, you shall be cast yourself into the Fire of Doom'. But Tolkien *was* an academic, and academics often see importance in academic issues where others do not. Is Frodo guilty? Has he given in to temptation? Or just been overpowered by evil? If one puts the questions like that, there is a surprising and ominous echo to them, which suggests that this whole debate between 'Boethian' and 'Manichaean' views, far from being one between orthodoxy and heresy, is at the absolute heart of the Christian religion itself. The Lord's Prayer, which in Tolkien's day everyone knew, and which most English-speakers know even yet, contains seven clauses or requests, and of these the sixth and seventh are:

> Lead us not into temptation,
> But deliver us from evil.

Are these variants of each other, saying the same thing? Or (much more likely) do they have different but complementary intentions, the first asking God to keep us safe from ourselves (the Boethian source of sin), the second asking for protection from outside (the source of evil in a Manichaean universe)? If the latter is the case, then Tolkien's double or ambiguous view of evil is not a flirtation with heresy after all, but expresses a truth about the nature of the universe denied to the philosopher Boethius, and possibly even to the rationalist Lewis.

There is no doubt that the Lord's Prayer was in Tolkien's mind as he wrote the Sammath Naur scene, for he said as much in a private letter to David Masson, with whom he had been discussing the criticisms made of him, as mentioned above. In this letter (kindly shown to me by Mr Masson, of the Brotherton Library in Leeds), Tolkien quoted the last three clauses of the Lord's Prayer, including 'Forgive us our trespasses', and commented that these were words which occurred to him, and that the scene in the Sammath Naur was meant to be a ' "fairy-story" exemplum' of them. Tolkien did not comment on the Prayer's apparent tautology, nor on the ambiguity of his own presentation of evil throughout, but they are of a piece. One can never tell for sure, in *The Lord of the Rings*, whether the danger of the Ring comes from inside, and is sinful, or from outside, and is merely hostile. And one has to say that this is one of the work's great strengths. We all recognize, in our better moments at least, that much harm comes from our own imperfections, sometimes terribly magnified, like traffic deaths from haste and aggression and reluctance to leave the party too soon: those are temptations. At the same time there are other disasters for which one feels no responsibility at all, like (as Tolkien was writing) bombs and gas-chambers. They may in fact all be connected, as Boethius insisted: no human being can ever see enough to tell. But our experience does not feel like that. It is a mistake just to blame everything on evil forces 'out there', the habit of xenophobes and popular journalists; just as much a mistake to luxuriate in self-analysis, the great skill of Tolkien's contemporaries, the cosseted upper-class writers of the 'modernist' movement.

And, of course, things would be much easier for the characters in *The Lord of the Rings* if this uncertainty over the nature of evil were to be withdrawn. If evil was just the absence of good, then the Ring could never be more than a psychic amplifier, and all the characters would need to do would be to put it aside, perhaps give it to Tom Bombadil: in Middle-earth we are assured that would be fatal. Conversely, if evil were only an external force without echo in the hearts of the good, then someone might have to take it to Orodruin, but it would not need to be Frodo: Gandalf could take it, or Galadriel, and whoever did so would have to fight only their enemies, not their friends or themselves. But if that were the case (and most fantasies are far more like that than *The Lord of the Rings*), then the work would be a lesser one, just a complex war-game of 'Dungeons and Dragons'; as it would be a lesser one if it veered instead in the direction of philosophical treatise or confessional novel, without relevance to the real world of war and politics from which Tolkien's experience of evil so clearly originated.

Positive forces: 1
Luck

One more question about the scene in the Sammath Naur is, of course, what made Gollum fall? There is absolutely nothing in the text to say. It is just an accident: one more example of that 'biased fortune' which according to Colin Manlove makes Tolkien's work impossible to take seriously. But it is clearly not *just* an accident: it is the result of a string of decisions taken at one time or another. By Bilbo, who refrained from murdering Gollum when he had the chance many years before, in chapter 5 of *The Hobbit*. By Gandalf and the elves, who do not execute or dispose of him when he is in their power, but 'treat him with such kindness as they can find in their wise hearts'. By Frodo, who allows him to come along on the journey into Mordor, and even goes some way towards reforming him and returning him to being Sméagol again. Finally by Sam, who after many betrayals spares Gollum yet again on Mount Doom, clearly out of a sort of sympathy: the 'something that restrained him' from a killing thrust is an awareness of what it means to have borne the Ring. Gandalf hints prophetically at what will happen almost at the start of the story (in a passage probably written in late on). When Frodo says indignantly, 'what a pity that Bilbo did not stab that vile creature', Gandalf replies, picking up the implications of people's words as he so often does, 'it was Pity that stayed his hand'; furthermore Bilbo took so little harm from the addiction of the Ring, he asserts, 'because he began his ownership of the Ring so. With Pity.' Frodo remembers this conversation when he and Sam capture Gollum in the Dead Marshes, and he gets his reward for it, as Bilbo got his, with a kind of poetic justice. Frodo spares Gollum from Sting, and Gollum in the end rescues Frodo from the Ring. Gandalf's statement has been cited already, 'even the very wise cannot see all ends', but one can recognize now that it is a Boethian statement as well as one about narrative. Moreover, most people can see a pattern once it has been traced out, and the death of Gollum confirms a pattern already expressed several times in the proverbs of Middle-earth – 'a traitor may betray himself, and do good that he does not intend', for instance, which Gandalf again prophetically says to Pippin (V/4).

Of course, even if Gollum's death remains not just an accident, one may still feel the neatness of his fate to be 'too good to be true'. Before deciding about that, though, it is as well to consider the meaning of the terms 'fate' and 'accident', and others related to them. The word Tolkien uses is sometimes 'chance', but he tends to qualify it. Tom Bombadil says, when he rescues the hobbits from Willow-man, 'Just chance brought me then, *if chance you call it*' (my emphasis). Ruin was averted in the Northlands, Gandalf says in Appendix A (III), 'because I met Thorin Oakenshield one evening on the edge of Bree. A chance-meeting, *as we say in Middle-earth*' (my emphasis again). The suggestion both times is that 'chance' is just a word people use to explain things they do not understand, but that this is a sign only of the limits of their understanding. But if this is the case, what would a less limited understanding make of events?

There are a few hints in *The Lord of the Rings* of superhuman powers outside Middle-earth. We learn from *The Silmarillion* that Gandalf, like Saruman and indeed Sauron, is a Maia, a spiritual creature sent originally for the relief of humanity and the other sentient species. Indeed, Tolkien said in a letter addressed to Robert Murray in November 1954: 'I wd. venture to say that [Gandalf] was an incarnate "angel"' – surely meaning the word, as often, etymologically, i.e. as in Greek *angelos*, 'messenger'(see *Letters*, p. 202). Gandalf says himself that after the fight with the

Balrog, 'Naked I was sent back'. He does not say who sent him back, or who sent him in the first place, but we can infer again from *The Silmarillion* that it was the Valar, the powers protecting Middle-earth under God, or under Eru, the One. The Valar however show no sign of direct interference in the affairs of Middle-earth. The Gondorians shout 'May the Valar turn him aside!' as the oliphaunt charges, but we never know whether they do or not. The beast does swerve, but that could be chance again. Or is 'chance' the way the Valar work?

There are two things one can say about this. The first is that (as perhaps no one in the world knew better than Tolkien) people have a strong tendency to invent words which express their feeling *both* that some things are just accidents, *and* that there may well be some patterning force in just accidents. The Old English word is *wyrd*, which most glossaries and dictionaries translate as 'fate'. Tolkien knew that the etymologies of the two words were quite different, 'fate' coming from the Latin *fari*, 'to speak', so 'that which has been spoken', sc. by the gods. The Old English word derives from *weorþan*, 'to become': it means 'what has become, what's over', so among other things, 'history' – a historian is a *wyrdwritere*, a writer-down of *wyrd*. *Wyrd* can be an oppressive force, then, for no one can change the past; but it is perhaps not as oppressive as 'fate' or even 'fortune', which extend into the future. As mentioned above, however (p. 27), there is a curiously exact modern English parallel to it, the word 'luck'. The *OED* is reluctant to accept the idea, but it is an attractive thought that this derives from Old English *(ge)lingan*, 'to happen', and must have meant originally, 'what has happened, what has turned up': so you can have good luck or bad luck, depending as we say on 'the luck of the draw'. Almost everyone believes, however, even now, at some level or another, that luck means more than that. As was said in chapter I, the dwarvish belief that Bilbo is one of those people who possess more than their share of it has modern parallels. We no longer believe in 'the Fates', but we still personify luck, Lady Luck; people say someone is having 'a run of luck'; you can 'back your luck' – Farmer Giles does in Tolkien's story *Farmer Giles of Ham*, and it pays off; and you can give luck some assistance – Farmer Giles's advantageous position at the rear of the column when the dragon charges it is brought about 'As luck (or the grey mare herself) would have it'. 'Trusting to luck' *on its own* is not however thought to be a sensible strategy, and here again ancient and modern opinions are in striking agreement. '*Wyrd* often spares the man who is not doomed', says Beowulf, but he adds, 'as long as his courage holds'. Gimli says to Pippin and Merry, 'Luck served you there', but he too adds, 'but you seized your chance with both hands, one might say'. Pippin and Merry had indeed at that moment had some assistance from something, for as Gríshnakh drew his sword to kill them, an arrow pierced his hand: 'it was aimed with skill, or guided by fate'. But which?

As with the ironies of interlace, the logic of luck (or chance, or fate, or fortune, or accident, or even *wyrd*) seems in Tolkien's view to be this: there is no knowing how events will turn out, and it is certainly never a good idea for anyone to give up trying, whether out of despair or out of a passive confidence that some external power will intervene. If there is an external power (the Valar), it has to work through human or earthly agents, and if those agents give up, then the purpose of the external power will be thwarted. As Galadriel says, some of the things in her mirror 'never come to be'. One might note also that some power or other sent the dream that brought Boromir to Rivendell; but it sent the dream first and most often to Faramir, who was 'eager' to follow its warning, only to be brushed aside by his brother. Boromir says that he took over because 'the way was full of doubt and danger', but there is some reason to disbelieve him. It would probably have been better for all if Faramir had been allowed to take the Valar's advice. But people can avert the intentions of Providence, and obeying them (in so far as they can be detected) brings no guarantee of success or safety. The most one can say is that luck may turn out better than one expects, as in the case of Gollum in the Sammath Naur: but your courage has to hold (so Beowulf), you have to seize your opportunity with both hands (so Gimli), and being 'too eager to deal out death in judgement', and more generally knowingly doing wrong to improve your chances, will probably be counter-productive (so Gandalf). Only the last opinion is really open to the charge of bias, and of that no mere mortal can be sure.

Positive forces: 2
Courage

Tolkien was, to put it mildly, not fortunate in his critics. They accused him of rigging the plot – I have tried to answer that just above. They accused him of failing to obey his own ground-rules over the Ring – I have tried to answer that by the word 'addiction'. They accused him of making his good and evil characters morally indistinguishable: this was answered with fierce logic by W.H. Auden, who pointed out first in 1955, then in 1961, that a major difference was that the good characters, the Gandalfs and the Galadriels, could imagine becoming bad, whereas Sauron's great weakness, even tactically, is that he cannot imagine the self-destructive strategy of destroying the Ring for ever. Just to show that any stick would do to beat some dogs, other critics complained that the good characters were just too good, without the expected human admixture of sin and weakness – thus taking no notice of the wraiths and the whole consistent idea of the wraithing process. However, one complaint which particularly annoyed Tolkien was that by Edwin Muir in the *Observer* (see *Letters*, p. 230). Muir reviewed each volume of *The Lord of the Rings* as they came out, successively in the *Observer* for 22nd August 1954, 21st November 1954, and 27th November 1955, and especially on the first and third occasions with strong reservations (he did like the Ents). Muir's complaint in the third review, the annoying one, was that the whole work was sub-adult in its painlessness: 'The good boys, having fought a deadly battle, emerge at the end of it well, triumphant and happy, as boys would naturally expect to do.' There is a simple reply to this, which is to say that Frodo at least does not end up well, or happy, and that he avoids any suggestion of triumph, seeming in the end incurably scarred, a 'burnt-out case'. He is admittedly taken away to be healed of his wounds, like King Arthur, though that is not the way Muir put it. But there are other people, and creatures, and things, which cannot be taken away or healed. In fact it is much easier to make a case out for Tolkien as a pessimist than as a foolish or childish optimist; it is another of the qualities which mark him out from most of those who have imitated him.

Thus, it is obvious that many if not most of the senior characters in *The Lord of the Rings* envisage defeat as a long-term prospect. Galadriel says, 'Through ages of the world we have fought the long defeat'. Elrond agrees, saying 'I have seen three ages in the West of the world, and many defeats and many fruitless victories'. Later on he queries his own adjective 'fruitless', but still repeats that the victory long ago in which Sauron was overthrown but not destroyed 'did not achieve its end'. The whole history of Middle-earth seems to show that good is attained only at vast expense while evil recuperates almost at will. Thangorodrim is broken without evil being 'broken for ever', as the elves had expected. Númenor is drowned without getting rid of Sauron. Sauron is defeated and his Ring taken by Isildur, but this only sets in motion the crisis at the end of the Third Age. Moreover, it is made extremely (one might have thought, unmistakably) clear that even the destruction of the Ring and the overthrow of Sauron will conform to the general pattern of 'fruitlessness' – or perhaps one should say that the fruit will be bitter. The destruction of the Ring, says Galadriel, will mean that her ring and

Gandalf's and Elrond's will all lose their power, so that Lothlórien 'fades' and the elves 'dwindle'. Along with them will go the ents and the dwarves, indeed the whole of Middle-earth, to be replaced by modernity and the dominion of men; all the characters and their story will shrink to misunderstood words in poems here and there, lists of names with their meaning forgotten like the *Dvergatal*, correspondences visible only to the philologist. Beauty especially will be a casualty. Théoden asks in 'The Road to Isengard' (III/8), 'However the fortunes of war shall go, may it not so end that much that was fair and wonderful shall pass out of Middle-earth?' Gandalf replies only, 'The evil of Sauron cannot be wholly cured, nor made as if it had not been.' Treebeard confirms Théoden's fear in much the same way as Gandalf when he says of his own doomed and dying species, 'songs like trees bear fruit only in their own time and their own way, and sometimes they are withered untimely' (III/4). The collective opinion of Middle-earth might well be summed up in Gandalf's aphoristic statement, 'I am Gandalf, Gandalf the White, but Black is mightier still' (III/5).

This sounds ominously like a Manichaean statement, and also a 'defeatist' one. However, as has been said, Tolkien was careful to voice rebuttals of Manichaeanism several times and in several ways. With his best friends dead in Flanders, he was likely also to have no patience whatsoever with 'defeatism' in its original meaning, French *défaitisme*, a word which came into being about 1918 to express the war-weariness of the Allies, the feeling (especially among civilians) that the sacrifices already made should now be abandoned for an inconclusive peace. Why then the work's continuing pessimism and expectations of defeat?

One answer must be that Tolkien wanted in a way to re-introduce to the world 'the theory of courage': not just courage, N.B., nor images of courage, but the 'theory of courage', which he had said in his *Beowulf* lecture of 1936 was the 'great contribution' to humanity of the old literature of the North (*Essays,*

p. 20). What Tolkien meant was this. The mythology we find still expressed in Old Norse (Tolkien believed that it must have been present also, and earlier on, in Old English) was like the traditional Christian one in that it too ended in a Day of Doom, an Armageddon, in which the forces of good and evil finally confronted each other. The difference was that in the Norse one it was the forces of *evil*, the giants and the monsters, which won, so that the Norse Armageddon was called Ragnarök, 'the destruction of the gods'. If the gods and their human allies are going to lose, though, and this is known to everyone, what in the world would make anyone want to join that side? Why not imitate the monsters instead, or become, so to speak, a devil-worshipper? The truly courageous answer – Tolkien called it a 'potent but terrible solution' (*Essays*, p. 26) – is to say that victory or defeat have nothing to do with right and wrong, and that even if the universe is controlled beyond redemption by hostile and evil forces, that is not enough to make a hero change sides. In a sense this Northern mythology asks more of people than Christianity does, for it offers them no heaven, no salvation, no reward for virtue except the sombre satisfaction of having done right. Even the heathen Valhalla is only a waiting-room and training-ground for the final defeat. Tolkien wanted his characters in *The Lord of the Rings* to live up to the same high standard, and was careful therefore to remove easy hope from them, to make them conscious of long-term defeat and doom.

Nevertheless Tolkien was himself a Christian, who did not believe that the universe was controlled 'beyond redemption' by the powers of evil; and he lived in a world in which the 'potent and terrible solution' of the 'theory of courage' had vanished almost beyond revival, or even understanding (try fitting it, for instance, into the plot of *Star Wars*). In his academic work he became accordingly significantly more nervous about seeing continuity from pagan to Christian eras in Old English poems – as one can see from his 1953 essay-poem rewriting *The Battle of*

Maldon as 'The Homecoming of Beorhtnoth Beorhthelm's Son', see pp. 294–6 below. In his creative work he needed a new image for ultimate bravery, one which would have some meaning and some hope of emulation for the modern and un- or anti-heroic world. It was a problem he had faced before (see chapter I, and the discussion of Bilbo's modernistic style of courage as opposed to the traditional heroic model of Beorn or Thorin); and in *The Lord of the Rings* he was to solve it once again through the hobbits, in a development of the cold-blooded, solitary, and non-aggressive courage shown in *The Hobbit* by Bilbo. But this time the hobbits are not so solitary, usually found in pairs, and the image of courage they project has a more social element about it: it is centred, unexpectedly, on laughter, cheerfulness, an attitude which far from speculating about its chances on Doomsday refuses ever to look into the future at all. There is about it at times an element of deliberate paradox.

All four hobbits in the main story are militarily unambitious, even in their most dramatic moments. Merry stabs the Nazgûl when the 'slow-kindled courage' of his race finally stirs, but it is from behind; Pippin is allowed to stab the hill-troll, one might think, in order to let him 'draw level with old Merry'. They seem, however, to be less affected than most of their seniors by the despair and demoralization which is the main weapon of the Black Riders, perhaps because they are less sensitive, perhaps because they refuse to predict or rationalize. In Minas Tirith it is Pippin who cheers Beregond when they hear the Rider's cry the first time, pointing to the sun and the banners and declaring, 'my heart will not yet despair'. Merry's duty meanwhile is to 'lighten [Théoden's] heart with tales'. He feels the 'great weight of horror and doubt' which settles on him and the Riders at the end of 'The Ride of the Rohirrim', but he is also the first to see the change. Part of the reason may be their (highly English) frivolity. They joke continually with each other, and with Théoden – who being English as well takes it in the right spirit. Merry apologizes for the habit in 'The Houses of Healing', 'it is the way of my people to use light words at such times', but Pippin's last thought, as he falls under the weight of the troll:

> laughed a little within him ere it fled, almost gay it seemed to be casting off at last all doubt and care and fear.

Pippin at that moment thinks he is dead and his cause totally defeated, but what cheers him at this last moment is the thought that he has been right all along, 'it ends as I guessed it would'. Sam and Frodo react in much the same way after the destruction of the Ring. They too think that they are as good as dead, and furthermore that this is exactly what one might have expected. Frodo indeed brings up but rejects the idea of the compulsory happy ending, saying seriously, 'it's like things are in the world. Hopes fail. An end comes ... We are lost in ruin and downfall, and there is no escape'. He turns out to be wrong, of course, but there is no denying the force of what he says *as a general rule*. But Sam's reaction is merely to reflect, 'What a tale we have been in', and to wonder what the title of it will eventually be. Sam in any case had reached the point of paradox rather earlier (IV/3), in that, Tolkien explains, he had:

> never had any real hope in the affair from the beginning; but being a cheerful hobbit he had not needed hope, as long as despair could be postponed. Now they had come to the bitter end. But he had stuck to his master all the way; that was what he had chiefly come for, and he would still stick to him.

Is it possible, one might ask, to be 'cheerful' and without hope at the same time? Modern optimistic convention says not ('Ya gotta have hope', says the song in the American musical), but the Gamgee family seems to take a sceptical view of that idea:

While there's life there's hope, says the Gaffer, conventionally enough, but he usually tacks on the deflating words, *and need of vittles*. Sam is in a way presenting a modern version of the 'theory of courage', which did not have to be offered the bribe of assured victory at Ragnarök to do its duty. Perhaps the argument may be that only those who need hope to keep going will fall prey to despair when their hope is withdrawn. Those who, like Sam and Pippin, felt from the start that the whole thing was going to be a disaster remain immune, even cheerful, when their expectations are confirmed. Tolkien knew that in the Norse mythology Vön, Hope, is not one of the three cardinal virtues but, contemptuously, the drool that runs from the mouth of Fenris-Wolf; he also knew that 'cheerfulness' is in its origin at least a virtue of the face alone, *chair* being the Old French word for face – when Sir Gawain sets off to face apparently certain death, the poet remarks (Tolkien's translation):

> The knight ever made good cheer [= put a good face on it],
> saying, 'Why should I be dismayed?
> Of doom the fair or drear
> by a man must be assayed.'

Modern convention again disagrees, but there is an old and still-powerful opinion which says that the face is more important than the heart: because it is, or it should be, under conscious control.

The most characteristic moment of Tolkien's new-model theory of courage is however at the end of Book IV/8, 'The Stairs of Cirith Ungol'. Here Sam and Frodo take what they expect to be their last meal, and consider the theory of narrative. The great tales do not come to an end, says Frodo, but in case that should sound too optimistic he adds that the people in them do. Sam pursues the idea of the continuing tale, and suggests with comic loss of grammar, and indeed bathos, that maybe in the future some father will say to his son that Frodo was 'the famousest of the hobbits, and that's saying a lot' (actually, we know, not much). And Frodo laughs:

> Such a sound had not been heard in those places since Sauron came to Middle-earth. To Sam suddenly it seemed as if all the stones were listening and the tall rocks leaning over them. But Frodo did not heed them; he laughed again. 'Why, Sam,' he said, 'to hear you somehow makes me as merry as if the story was already written. But you've left out one of the chief characters: Samwise the stouthearted. "I want to hear more about Sam, dad. Why didn't they put in more of his talk, dad? That's what I like, it makes me laugh."'

They carry on talking, and then fall asleep. There Gollum finds them, to be touched by the peace in their faces, and to creep up and try to touch Frodo's knee, seeming for a moment not Gollum but Sméagol again, 'an old weary hobbit, shrunken by the years . . . an old starved pitiable thing'.

It is a sign of a kind of hardness in the fable, not recognized by Edwin Muir, that this moment of sentiment is immediately dissipated by Sam. He wakes up, sees Gollum ' "pawing at master," as he thought', speaks to him roughly (Gollum for once answers 'softly'), and then accuses him of 'sneaking off and sneaking back – you old villain'. At this 'Gollum withdrew himself . . . The fleeting moment had passed, beyond recall'. Among the unnoticed casualties of Middle-earth, one should realize, is the old hobbit Sméagol, as well as the creature he turns into, Gollum. Most of the characters indeed bear a burden of regret, expressed again with deliberate paradox in Treebeard, who knows his race and his story are sterile, but who looks as a result, according to Pippin, 'sad but not unhappy' (III/4). Can you be 'sad' and '(not un)happy' at the same time? Not according to modern semantics, but Tolkien often took no notice of that. Treebeard's sad happiness (an old meaning of 'sad' is 'settled, determined', as C.S. Lewis pointed out in his *Studies on Words*), and Sam's hopeless cheer (identical with Sir Gawain's), form an image of courage which, above all, can carry on in the complete absence of any trust in luck.

Some conclusions

There is a final unrecognized touch of hardness in *The Lord of the Rings* in the oblivion that is settling over Frodo in its final chapters. He may have great honour elsewhere, but in the Shire Sam is 'pained' to see how little respect he is given. His pacifism and lack of aggression mean that he takes no part in the final 'Battle of Bywater' except to intervene to protect the prisoners. His name is not at the top of the Roll which has to be learned by heart by all Shire-historians, and his family, unlike the Cottons, Gamgees, and Fairbairns, gains no advancement from it. Indeed he has no family: his story, like Treebeard's, will prove to be sterile, and in spite of Sam's imaginings he will never be 'the famousest of the hobbits' in the Shire itself. Nor does he appear to be curable, at any rate in this world. And he insists on almost the last page of the entire work, as he and others have done all the way through, that that is the way things are:

> 'It must often be so, Sam, when things are in danger: some one has to give them up, lose them, so that others may keep them.'

The lack of respect and attention paid to this best of the 'good boys' contradicts not only Edwin Muir but the whole system of war memorials, minutes' silences, Earl Haig funds and poppy-days familiar to Tolkien as to all inhabitants of post-World War I Britain. Or at least it does so on the surface: what Frodo says is in fact similar to the words on the Imphal-Kohima monument, now itself largely forgotten, and themselves often misquoted. They go (in evident parallel to the epitaph of Simonides for the Spartans at Thermopylae, Golding's 'Hot Gates'):

> When you go home tell them of us and say
> For your tomorrow we gave our today.

Like several of the major authors of fantasy mentioned above, Tolkien was a war-survivor, and his work expresses along with a strong belief in (a kind of) Providence the disillusionment of the returned veteran.

If one turns now from the incipiently sublime to the near-ridiculous, there is, as I have suggested before, a fierce and a strong competition among literary critics for the honour of having made the least perceptive comment on Tolkien. One of the contenders must be the dismissal offered by Professor Mark Roberts of Keele University, who said of *The Lord of the Rings* in an article in *Essays in Criticism* for 1956:

> It doesn't issue from an understanding of reality which is not to be denied, it is not moulded by some controlling vision of things which is at the same time its *raison d'être*.

In this post-modern world it is of course hard to conceive of any 'understanding of reality' which will not be denied by someone or other, but Professor Roberts spoke from a simpler critical era; he was clearly trying to write Tolkien off in the language and from the perspective of the then-dominant F.R. Leavis – and indeed it is true that *The Lord of the Rings*, like the rest of modern fantasy, would never fit into the neat succession of Leavis's 'Great Tradition'. When Roberts says, however, that the work has no 'controlling vision of things which is at the same time its *raison d' être*', one has to wonder how he could be so blind. As I have tried to show in this chapter and the preceding one, *The Lord of the Rings* fits together, whether one likes the result or not, on almost every level. The complex interlacement of the narrative structure positively generates ironies (and anti-ironies) for the reader, uncertainties and 'bewilderment' for the characters. Those uncertainties, about themselves and others, are mirrored by the ambiguous nature of the Ring, part psychic amplifier, part malign power, and of the ultimate source of the evil that surrounds it, perhaps internal, perhaps external. I have argued that the work's 'controlling vision of things' is in fact a double vision, between the opinions I label 'Boethian' and 'Manichaean'; and that both opinions are presented at one time or another with equal force, whether it is in the Dead Marshes (Manichaean, but maybe an illusion) or the Field of Cormallen (Boethian, but rapidly evanescent). As the characters steer their way through these consistent uncertainties, they are guided by a developed theory of 'chance' or 'luck' which is at the same time perfectly familiar, perfectly colloquial, and also philologically and philosophically consistent; and by a theory of courage which is similarly ancient in its roots, and familiar in contemporary times (as Tolkien said himself) from one First World War memoir after another. It is reasonable to imagine someone rejecting Tolkien's vision (though it has proved powerful to many, like me, who unlike him are not committed Christians). But there is something wilful and weary about the inability to see that there is any vision there at all.

What Professor Roberts meant, no doubt, was that Tolkien did not share his vision and the vision of his class and time, and the difference is especially strong over the question of the nature and source of evil, which I take to be the central issue of *The Lord of the Rings*, as of so many modern fantasies. It is worth reflecting for a moment on the opinions about this available to Tolkien and his fellow-veterans from the official spokesmen and spokeswomen of his contemporary culture, say in the 1920s and

1930s. There was a Freudian view, slowly making its way into general consciousness in the early years of the twentieth century, as one can see from the *OED*'s reluctant and belated entries on words like 'repression', 'complex', 'unconscious', 'trauma': Lewis in particular continually objected to this, probably with the backing of his fellow-Inklings, as tending to dissolve responsibility or any sense of personal guilt. There was what one might call the 'Bloomsbury' view, expressed by writers like Virginia Woolf, E.M. Forster, Bertrand Russell, and above all G.E. Moore, whose *Principia Ethica* has been called the most significant philosophical work of the century. But while summarizing the views of all the 'Bloomsberries' together would be a difficult occupation, one can say with some conviction that there is nothing in the *Principia Ethica* which has any but the remotest bearing on the immediate issues of evil in the twentieth century – industrialized war, carpet bombing, the use of chemical, biological and nuclear weapons, genocide and the massacre of non-combatants, several of these matters of personal experience to Tolkien and his fellow-veterans. Bloomsbury views of vice and virtue are essentially private; they are like Freud's in being above all about human relationships.

Meanwhile, and even more prominent than the last two, there are the traditional views and images of evil as put forward by literary tradition. It is astonishing how often Tolkien was reprimanded by critics for not returning to these images – which, however, his fellow-writers (often and for good reason keen medievalists) had seen, considered, and discarded as no longer possible, 'not *enough* any more', to repeat Vonnegut on Dostoyevsky. Why could Tolkien not be more like Sir Thomas Malory, asked Muir, in the third *Observer* review of those cited above, and give us heroes and heroines like Lancelot and Guinevere, who 'knew temptation, were sometimes unfaithful to their vows', were engagingly marked by adulterous passion? But T.H. White had already considered that paradigm, was indeed rewriting it at the same time as Tolkien as *The Once and Future King*; and he had seen the core of Malory's work not in romantic vice but in the human urge to murder. In White the poisonous adder which allegedly provokes the last disastrous battle is no adder but a harmless grass-snake, and the flash of the sword which brings on the two armies is not natural self-defence but natural blood-lust, creating a continuum from cruelty to animals to world wars and holocausts. Malory has to be rewritten to encompass a new view of evil. Or, Muir asked in the first of his reviews, why could Tolkien have not given us anti-heroes more like the Satan of *Paradise Lost*, at once 'both evil and tragic'? But C.S. Lewis had already considered that paradigm, had indeed rewritten *Paradise Lost* in 1943 as *Perelandra*, or *Voyage to Venus*. In that he made clear his opinion that there was nothing at all grand, dignified and tragic about evil, which was instead tedious, sordid and squalid, showing itself in petty mutilations which disgusted even a man like his hero Ransom who had been (like Tolkien) 'on the Somme'. Other writers, like Vonnegut and Heller, spent close to twenty years trying to work out a literary mode which could encompass their experiences of insanity, the absurd, the non-volitional. The 'Great Tradition' of the English novel was no use to them in this.

To return to what was said at the beginning of this chapter, for all its antiquarian knowledge and antiquarian charm, no one could mistake *The Lord of the Rings* for anything but a work of the twentieth century. It shows above all the difficulties which that century has created for traditional views of good and evil, though it also tries to re-assert them. Aragorn says to Éomer, 'Good and ill have not changed since yesteryear; nor are they one thing among Elves and Dwarves and another among Men. It is a man's part to discern them.' But one may feel as much sympathy with Éomer's uncertainty, and with his question, 'How shall a man judge what to do in such times?' In the next chapter I go on to discuss the contemporaneity of *The Lord of the Rings*, which has led some into thoughts of political allegory, and also the drive towards a more enduring mythology which I believe underlies contemporary applications.

CHAPTER IV

THE LORD OF THE RINGS (3):
THE MYTHIC DIMENSION

Allegory and applicability

In the 'Foreword' to the second edition of *The Lord of the Rings*, Tolkien wrote: 'I cordially dislike allegory in all its manifestations, and always have done since I grew old and wary enough to detect its presence'. As with the denial of any link between rabbits and hobbits (see chapter I), the evidence is rather against Tolkien here. He was perfectly capable of using allegory himself, and did so several times in his academic works, usually with devastating effect. In his 1936 lecture on *Beowulf*, for instance, Tolkien offered his British Academy audience 'yet another allegory' (it was not the first in the lecture), about a *man* who built a *tower*. He took the *stone* for the tower from a *ruin*, 'an accumulation of old stone' in a field, part of which had also been used to build the *house* in which the man actually lived, 'not far from the old house of his fathers' (i.e. the ruin). But his *friends* came along, noticed at once that the tower was made of older stones, and laboriously knocked the tower down to examine the stones, look for carvings on them, prospect for coal, and so on. Then some of them complained that the tower was in a terrible mess, while even the man's *descendants* murmured that he should have spent his time not building the tower but restoring the ruin. 'But from the top of that tower the man had been able to look out upon the sea' (see *Essays*, pp. 7–8).

There is no doubt that this *is* an allegory, for Tolkien says so himself. A brief study of it, concentrating on the elements italicized above, may explain exactly what Tolkien meant by the word, how he expected allegories to work, and why he disliked both word and thing when they were misused. Tolkien's little story is an allegory of the progress of *Beowulf* criticism, one of the major features of which, up to Tolkien's time, had been a conviction that the poet had written the wrong poem. Its accuracy, or 'justness' to use Tolkien's own term, is hard to appreciate without the kind of awareness of *Beowulf* scholarship which Tolkien's original audience may be supposed to have had, but in brief one could say that:

The *old stone*, i.e. the *ruin* = the remains of an earlier, heathen, oral poetry which the *Beowulf*-poet might have known about

The *house* the man lives in, also partly built from the ruin = Christian poetry contemporary with *Beowulf* like the poem *Exodus* (Tolkien's edition of which was published posthumously in 1981), which also drew on the early oral poetry

The *tower*, of course = *Beowulf*, and the *man* = the *Beowulf*-poet

The man's *friends* who knock his tower down = the dissectionist critics of the nineteenth century, who concentrated their efforts on pointing out where the poem had gone wrong

Finally, the man's *descendants*, who wished he had restored the old house = British critics like W.P. Ker and R.W. Chambers, who rejected dissectionism but said repeatedly that they wished the poet had written an epic about history rather than a mere fairy-tale about dragons and monsters.

The main point about the above, though, is the repeated = sign. Tolkien did not think that allegories made sense unless you could consistently and without error fill these in. And to him the function of allegory was usually, as in this case, as a *reductio ad absurdum*. Anyone listening to Tolkien's allegory of the tower would sympathize with the tower-builder, and not with the short-sighted fools who destroyed it. Therefore, Tolkien implied, they should sympathize with the poem and not with its critics.

That was why Tolkien, in the 'Foreword', dismissed contemptuously those who would see *The Lord of the Rings* as an allegory of World War II. In the first place, as he pointed out, he started work on it 'long before the foreshadow of 1939 had yet become a threat of inevitable disaster'. But in the second place the 'equals' signs were missing. One could, of course, say that the Ring = nuclear weapons, the coalition of Rohan, Gondor and the Shire (etc.) = the Allied powers, Mordor = the Axis powers, all of which has some general plausibility. But in that case what does the destruction of the Ring and the refusal to use it equal? As Tolkien wrote in the 'Foreword', if this equation had been true, 'the Ring would have been seized and used against Sauron', as nuclear weapons were used against Japan; Barad-dûr would have been 'occupied', as the Axis powers were by the Allies; Sauron 'would not have been annihilated but enslaved'. As for Saruman, the unreliable ally, who would presumably have to 'equal' the USSR, he would 'in the confusion and treacheries of the time have found in Mordor the missing links in his own researches' and 'made a great Ring of his own', as the Russians used German scientists (Mordor) and Western agents (treachery) to make their own nuclear weapon. There *could* have been a Middle-earth allegory of World War II, Tolkien showed – but it would have been a quite different story, a significantly different story, from *The Lord of the Rings*.

One can accept, then, that Tolkien disliked vague allegories, allegories which didn't work, though he accepted them readily in their proper place, which was either advancing an argument (as in the *Beowulf* example) or else constructing brief and personal fables (like, in my opinion, some of his shorter pieces to be discussed in chapter 6). He was however prepared to accept something which might well look like allegory to the unskilled, as he also said in the 'Foreword'. Immediately following on from the sentence quoted at the start of this chapter, he wrote:

> I much prefer history, true or feigned, with its varied applicability to the thought and experience of readers. I think that many confuse 'applicability' with 'allegory'; but the one resides in the freedom of the reader, and the other in the purposed domination of the author.

To this he added further, 'An author cannot of course remain wholly unaffected by his experience'; but his experience, he reminded readers, probably went back further than theirs. 1914 was just as bad as 1939, if you were young then: 'By 1918 all but one of my close friends were dead'. And 'The Scouring of the Shire', with its felled trees and polluted rivers, reflected a process which went back long before the austerity years of the Labour government of 1945–50, so that the chapter 'had no allegorical significance or contemporary political reference whatsoever'. But that did not mean it meant nothing, and nor did the rejection of the World War II / nuclear weapons allegory mean that *The Lord of the Rings* had nothing at all to do with Tolkien's early twentieth-century experience.

Hints of correspondence between our history and the history of Middle-earth are in fact fairly frequent. Frodo says, when Gandalf tells him that the Shadow has returned to Mordor, 'I wish it need not have happened in my time', and Gandalf replies, 'so do all who live to see such times. But that is not for them to decide'. The phrase 'in my time' may recall Neville Chamberlain's now infamous promise, on his return from capitulating to Hitler in Munich in 1938, that he brought 'peace in our time'. He did not. Frodo's wish to put the whole thing off (not cure it) is as short-sighted as Chamberlain's turned out to be, and when Gandalf says 'that is not for them to decide', he is condemning the whole discredited idea of 'appeasement'. Much later Elrond, looking back on the past, remembers the time 'when Thangorodrim was broken, and the Elves deemed that evil was ended for ever, and it was not so'. The idea that evil could be ended 'for ever' may recall the belief that World War I was fought as 'the war to end all wars'; but Tolkien himself lived both through the time of that belief, or that assurance, and its total failure with the second outbreak of world war in 1939.

More significantly detailed, perhaps, is the side-issue of the Rammas Echor, which is what the men of Gondor call 'the out-wall that they had built with great labour, after Ithilien fell under the shadow of their Enemy'. This is first mentioned when Gandalf rides in to Minas Tirith with Pippin (V/1). It is still at this moment under construction, or under repair, and Gandalf tells the men working at it that they are 'over-late'. The wall is a waste of time, so 'leave your trowels and sharpen your swords!' They ignore him and carry on, and the next time the Rammas is mentioned, it is in council (V/4) when Denethor insists that the wall, 'made with so great a labour', cannot be abandoned. Faramir objects to the idea of garrisoning it, seconded by Imrahil, the Prince of Dol Amroth, but Denethor insists, and it is his insistence that leads to the wounding, and almost the death of Faramir later on in the same chapter. That apart, the Rammas Echor plays no real part in the story, and the vignette of Gandalf and the trowellers could have been left out without disturbance. On the other hand, the image of men and labour wasted guarding a wall which

J.R.R. TOLKIEN: AUTHOR OF THE CENTURY

served no purpose could hardly fail, in the 1950s, to remind readers of the Maginot Line, built (or half-built) in order to secure France for ever from German invasion, but in the end strategically pointless, influential only in contributing to a sense of false security. French experience is glanced at also in the encounter near the end of Book V with 'the Mouth of Sauron'. He presents the terms of Sauron's offer as follows, but I gloss them in the language of twentieth-century history: first, Gondor and its allies are to withdraw beyond the Anduin, taking oaths never again to attack (there will be a peace-treaty and an armistice); second, 'All lands east of the Anduin shall be Sauron's for ever, solely' (sovereignty over the disputed territory of Ithilien, the Alsace-Lorraine of Middle-earth, is to be transferred); third, there will be an area west of Anduin which 'shall be tributary to Mordor, and men there shall bear no weapons, but shall have leave to govern their own affairs. But they shall help to rebuild Isengard ... and that shall be Sauron's, and there his lieutenant shall dwell'. Gandalf and the others see through this last proposal straight away, but it is in effect the creation of a demilitarized zone, with what one can only call Vichy status, which will pay war-reparations, and be governed by what one can again only call a Quisling. 'Vichy' and 'Quisling' are words which were only proper names before the 1940s, without political meaning. As with Frodo's wish for postponement, which could become appeasement, they represent a natural urge to salvage something out of defeat, but an urge which Tolkien's own western world had learned by recent and bitter experience was even worse than the alternative.

Finally, in spite of Tolkien's own denials, one might wonder again about the 'applicability' of 'the Scouring of the Shire'. Tolkien said flatly that the chapter did not reflect 'the situation in England at the time when I was finishing my tale' (i.e. in the late 1940s); and for proof of this pointed out that it was an essential part of the plot, 'foreseen from the outset', and that the slender 'basis in experience' which it did have in fact dated much further back, to the time before World War II and indeed before World War I, when places like his old home in Sarehole, Warwickshire, were being drawn into the industrial Birmingham conurbation. Yet, as Tolkien also said, 'applicability . . . resides in the freedom of the reader', not 'the purposed domination of the author'. For most readers in the 1950s, as for anyone who still remembers the time, the homecoming chapters were bound to have points of resemblance with England – perhaps most of all because here and there they seem slightly out of place in Middle-earth.

Take, for instance, the matter of pipeweed. There no longer is any in the Shire, for Hob Hayward reports that 'All the stocks seem to have gone . . . waggon-loads of it went away down the old road out of the Southfarthing'. One wonders where to. Saruman has certainly become a smoker, which may explain the small amounts being traded earlier on, but he cannot be smoking 'waggon-loads' of it himself, and it seems hard to believe that he is either selling it or issuing it to his followers in Isengard. In the 1940s and 1950s, though, it was common in Britain for shortages to be explained with a shrug and the words 'gone for export'; this created exactly the annoying paradox of the Shire, of plenty of production but no consumption and no other visible benefit. More significant may be the curious 'socialism', so to speak, of Sharkey and his men. They are robbers and bandits, and Sharkey/ Saruman's only goal is vengeance, as he says himself, but it is strange that the ruffians camouflage their intentions with some sort of ethic of fairness. 'It's all these "gatherers" and "sharers"', says Hob Hayward again, 'going round counting and measuring and taking off to storage. They do more gathering than sharing, and we never see most of the stuff again'. Farmer Cotton confirms that they gather stuff up 'for fair distribution', and the quotation marks indicate that that is their phrase, not his; he also admits that some fraction of this does come back, as the 'leavings' at the Shirriff-houses.

All this seems oddly euphemistic for Middle-earth, whose villains usually feel no need to conceal their intentions. But Sharkey's men do actually seem to believe their own rhetoric themselves. 'This country wants waking up and setting to rights', says the leader of the Hobbiton ruffians, as though he had some goal beyond mere hatred and contempt for the Shire, and in so far as this ever becomes visible, it seems to be more industrialization, efficiency, economy of effort, all things often and still wished on the population of Britain. The trouble with that (as developments after the publication of *The Lord of the Rings* have tended to confirm) was that the products of efficiency-drives were often not only soulless but also inefficient. Why do Sharkey's men knock down perfectly satisfactory old houses and put up in their place damp, ugly, badly-built, standardized ones? No one ever explains, but the overall picture was one all too familiar to postwar Britons, as was the disillusionment of returning from victory to poor food, ration-books, endemic shortages, and a rash of 'prefabs' and jerry-built 'council houses'. So was the suspicion that behind every 'Lotho Pimple' or domestic tyrant, there was some more sinister force, which would eventually take over and indeed devour the wretched Lothos. In 'The Scouring of the Shire' one can be sure that there is a good deal of Tolkien's own early experience and personal feelings, especially over the loss of trees; though one can also see why he would not want these late chapters to be taken as a mere allegory of or attack on the Socialist government of Britain 1945–50, for that would be a petty and a transient conclusion for a work of such scope. Nevertheless, just as the Rammas Echor sounds a warning against the wish for safe solutions which bred the Maginot Line, so 'The Scouring of the Shire' gives a reminder that the loss and damage of wars do not end with the victory parades, but drag on in the drabness and poverty which Orwell (writing at exactly the same time) projected into the future as the 'Ingsoc' of *Nineteen Eighty-Four*.

Saruman and Denethor: technologist and reactionary

There is a strong applicability, furthermore, in the characters Saruman and Denethor. One of the things that connect them is this. As I have said many times already, one of Tolkien's characteristic activities was the antiquarian one of showing old words and beliefs and habits, like the riddle-game, persisting in to modern times. But he was also capable of working the other way: taking something apparently distinctively modern, and wondering what it would have been like in the different circumstances of an archaic world. Is the modern element really modern? Or was it there all the time, unnoticed, waiting to be revealed?

These questions apply with particular force to Saruman. One should note to begin with that his name, as so often, comes from a philological puzzle, though it is one which no one had identified before Tolkien, and to which he gave a highly personalized solution. The puzzle is what the word *searu* might mean in Old English (*searu* is the recorded West Saxon form, **saru* would be its unrecorded equivalent in Mercian, the language of the Mark, Tolkien's West Midland home). This word survives in several compounds, and it has associations first with metal – Beowulf's mail shirt is a *searonet*, a *searo*-net, sewed, we are told, by the cunning thought (*orþanc*) of the smith. The Danish hall of Heorot is held together by *searo*-bonds, presumably iron clamps. The standard translation for *searo* here is something like 'cunning', and this fits with other uses, such as the description of the thoughts of wizards as *searoponc*, 'cunning thought'. The word has ominous suggestions as well, in the adjective *searocræftig* or the noun *searoniþ*, 'cunning-crafty', 'cunning-spite'. Finally it is connected also with treasure, which may be a *searugimma geþræc*, a 'confusion of cunning gems'; in one final crux, one Old English poem declares, making the word into a verb, *sinc searwade* – 'treasure was cunning'. The standard dictionary suggests that this means it 'left its possessor', but Tolkien was more likely to think it meant the exact opposite, treasure *stayed with* its possessor, gave him the 'dragon-sickness' of which the Master of Laketown died.

Saruman could then mean simply 'cunning man', itself an old designation for a wizard, and so suitable enough. But behind that one may see that for Tolkien the Old English word expressed very accurately a complex concept for which we no longer have a term. What does Saruman stand for? One thing, certainly, is a kind of mechanical ingenuity, smithcraft developed into engineering skills. Treebeard says of Saruman that 'He has a mind of metal and wheels'; his orcs use a kind of gunpowder at Helm's Deep, and later on he uses against the Ents a kind of napalm, or (remembering Tolkien's own war experience) perhaps one should say a *Flammenwerfer*. This might in itself be ethically neutral, and Tolkien had a lifelong sympathy with all kinds of creative endeavour, including the forging of the Silmarils and 'the love of beautiful things made by hands and by cunning and by magic', the 'desire of the hearts of dwarves' which Bilbo feels for a moment in chapter 1 of *The Hobbit*. Yet in the dwarves, just as in the Old English poem mentioned above, this love can lead on to something greedier and more treacherous, the 'bewilderment of the treasure' which works on Thorin as on the Master of Laketown. There is another connection entirely personal to Tolkien. In his childhood he lived in the village outside Birmingham then called 'Sarehole', and feared the bone-grinding millers he and his brother called 'the White Ogre' and 'the Black Ogre' (see *Biography*, pp. 20–21). Sarehole Mill became for him an image of destructive technology, remembered in the scenes with the miller Ted Sandyman in the Shire. How suitable that 'Sarehole' could be taken to mean 'the *saru*-pit' or possibly 'the sere pit, the withered pit'. Ted Sandyman is withered, in a way, says Farmer Cotton (VI/8): 'he works there cleaning wheels for the Men, where his dad was the Miller and his own master'. His is not the dragon-sickness associated with gold, but it is a metal-sickness associated with iron, and both sicknesses are catching and potentially fatal.

One sees 'Sandyman's disease' in an advanced form in Saruman: it starts as intellectual curiosity, develops as engineering skill, turns into greed and the desire to dominate, corrupts further into a hatred and contempt of the natural world which goes beyond any rational desire to use it. Saruman's orcs start by felling trees for the furnaces, but they end up felling them for fun, as Treebeard complains (III/4). The 'applicability' of this is obvious, with Saruman becoming an image of one of the charac-teristic vices of modernity, though we still have no name for it – a kind of restless ingenuity, skill without purpose, bulldozing for the sake of change. It is interesting that Saruman's followers call him 'Sharkey', Orkish, we are told, for 'Old Man'. A medieval-ist might think of the 'Old Man of the Mountains' in Sir John Mandeville's *Travels*, the master of the sect of Assassins. His title in Arabic would be *shaikh*, 'old man', and he ruled by deluding his followers with dreams of Paradise created by *hashish*. Simi-larly, one might say, the Sarumans of the real world rule by deluding their followers with images of a technological Paradise in the future, a modernist Utopia; but what one often gets (and this has become only more relevant since Tolkien wrote and since he died) are the blasted landscapes of Eastern Europe, strip-mined, polluted, and even radioactive. One may disagree with Tolkien's diagnosis of the situation, and with his nostalgic or pastoral solution to it, but there can be no doubt that he has at least addressed a serious issue, and tried to give it both a historical and a psychological dimension nearly always missing elsewhere.

In view of the 'socialist' suggestions clinging to Saruman, one ought to point out that he has a counterpart who is very clearly an arch-'conservative'. This is Denethor. Denethor's most evident counterpart is Théoden, as has been said, both old men who have lost their sons, representatives of cultures clearly contrasted to one another. Their fates are also both connected and contrasted.

Both men at one time or another succumb to despair, both are encouraged by Gandalf, each gains a hobbit-squire who attempts to cheer him further. At a critical moment, however, they react differently. At the end of 'The Ride of the Rohirrim' Merry sees Théoden seemingly quail before the Black Breath, the 'great weight of horror and doubt', and fears that he will turn back, 'slink away to hide in the hills'. Instead he cries out five lines of heroic verse, an extension of the 'call to arms' he had given on taking Éomer's sword in 'The King of the Golden Hall', blows the horn he takes from his banner-bearer, and charges. At exactly the same moment, carefully synchronized for us by the horn-blowing heard in Minas Tirith, Denethor is attempting to commit suicide and to take his son Faramir with him. In yet another example of the close cross-referencing of interlace, we know exactly why, and what mistake Denethor has made. The key reference is some thirty pages before, when, after Faramir has been brought in wounded, Denethor retires to his secret room, and watchers see 'a pale light' gleaming and flickering from the windows. Denethor is looking in a *palantír*, as is confirmed later by Gandalf and Beregond. In it he sees, for one thing, the black sails of the Corsairs coming up the Anduin, but that must be on the day of the battle itself, March 15th, or the day before, March 14th. What did he see on the 13th, the day when Faramir was brought in, the day the 'pale light' was seen flickering? The 13th is the day when Frodo is captured and taken to Minas Morgul. The likelihood is that *that* is what Denethor has seen, in a vision controlled by Sauron. That is why he says to Pippin, speaking of Gandalf, 'The fool's hope has failed. The Enemy has found it, and now his power waxes'. *It* is the Ring, but (though the first-time reader does not yet know this) the Enemy has *not* found it. Indeed, the 15th, the day of the battle and the deaths of both Théoden and Denethor, is also the day when Frodo and Sam *escape*. Denethor's suicide is then ironically misguided, and in a further example of chains of cause and effect, Denethor kills not only himself but also his counterpart Théoden – for as said on p. 111 above, that is the implication of Gandalf's comment, as he hesitates before following Pippin to rescue Faramir, that 'if I do [save Faramir], then others will die'. He and Pippin hear the death-cry of the Lord of the Nazgûl as they come back from saving Faramir; while they have been doing so, Théoden has charged, met the Black Rider, been thrown, and is now saying his last words to Éomer and Merry.

Is there any 'applicability' in all this? One point surely is the difference between Théoden and Denethor, over all their similarities. Denethor is cleverer than Théoden, knows more, is more civilized and perhaps more intelligent: but he is not wiser. He does not understand about luck. He looks too far into the future, and misinterprets what he sees. Above all, he trusts to his own chains of logic. What he sees in the future is, in the first place, defeat, but even if that were to be avoided, change. He has worked out who Strider is, and sees that victory would mean being supplanted as Steward by 'the return of the King'. But he is not prepared to accept any change at all. When Gandalf asks him, almost at the last moment, what he wants, he says, 'I would have things as they were in all the days of my life . . . and in the days of my longfathers before me'. And if that is not possible, 'if doom denies this to me, then I will have *naught*: neither life diminished, nor love halved, nor honour abated'. By the time *The Lord of the Rings* was published, of course, it was for the first time in the history of the world possible for political leaders to say they would have 'naught', *and make it come true*. In this context there is a special ominousness in Denethor's prophecy, 'It shall all go up in a great fire, and all shall be ended. Ash! Ash and smoke blown away on the wind!' Denethor cannot say 'nuclear fire', but the thought fits. His decision to commit suicide and take his son with him rather than be enslaved has a kind of similarity to the doctrine of Mutual Assured Destruction, 'better dead than Red', which hung over the world from 1947 (the time of the first

Russian nuclear explosion). It is significant that Denethor's most repeated statement – he says it three times, once in V/4 and twice in V/7 – is 'The West has failed'. If Saruman suggests one kind of disastrous future familiar from twentieth-century experience, the technological Utopia turned into squalid dictatorship, Denethor represents the major late twentieth-century fear, leaders with a death-wish who have given up on conventional weapons. One sees how fortunate it is that Denethor did *not* gain control of the Ring.

None of this says, of course, that the Ring = nuclear weapons, the starting move of the would-be allegory which Tolkien mocked so decisively. Nevertheless some at least of the thoughts above, whether about the Maginot Line or council housing, Vichy status or the failure of the West, are bound to have occurred to any adult reader with any memory of recent history. They do not mean that *The Lord of the Rings* is a veiled rewrite of recent history, which would be an exercise with almost no point. They do mean that the patterns discernible in it, including the ironies of interlace and the moral they point out, can be applied to recent history and indeed to future action. The moral, obviously, is that one should never give up hope (like Denethor), nor on the other hand sit back and wait for things to change (like too many of the inhabitants of the Shire). But as Tolkien says, 'applicability ... resides in the freedom of the reader', and should only be suggested or provoked by the author.

Mythic mediation

A second area where one may feel inclined to disagree with Tolkien's overt statements about his own work is that of religious meaning. In a letter of 1953, written to a Jesuit friend, he claimed that:

The Lord of the Rings is of course a fundamentally religious and Catholic work; unconsciously so at first, but consciously in the revision. That is why I have not put in, or have cut out, practically all references to anything like 'religion', to cults or practices, in the imaginary world. For the religious element is absorbed into the story and the symbolism.

(*Letters*, p. 172)

The first thing one is bound to ask here is, what did he mean by 'fundamentally'? *The Lord of the Rings* is certainly superficially neither Catholic nor religious, nor Christian. As Tolkien says, there is almost no hint of any religious feeling at all in the characters or in their societies, not even where one would be most likely to expect it. The hobbits, for all their nineteenth-century Englishness, are devoid of any religious sanction for any of their activities. We know they get married, and might suppose that they did so rather like Tolkien's own contemporaries. But they have no churches, and there is no hint as to who marries them. The Mayor, in a 'civil ceremony'? The Thain? The Head Shirriff? The hobbits furthermore have elaborate genealogies, but apparently no tombstones or burying-grounds, whether in churchyards or out of them. The *dwarves* have tombstones, for we see that of Balin son of Fundin, but all we are told about them (in Appendix A (III)) is that the dwarves have many strange beliefs, apparently in a kind of reincarnation. The Riders again might be expected on general cultural grounds to have some sort of religion comparable to that of the ancient English before the time of their conversion to Christianity, and there are indeed hints of something like that in their past. On the border of Rohan is a mountain called the Halifirien, and this must be Old English *halig fyrgen*, 'Holy Mountain'. But we never find out who or what it was once holy to. They are also very scrupulous about burying their dead, both those killed in battle and their kings, all of them 'laid in mound', as Théoden is after the Battle of Pelennor. We get a detailed description of this in VI/6, with the Riders burying their king with his weapons and his treasure, raising a mound over him, and riding round his barrow singing a dirge. But what is remarkable here, perhaps, is what is missing. There is a description of a burial in real history very like Théoden's, that of Attila the Hun, and Tolkien refers to the text containing it in a letter to his son Christopher. In that description, though, the barbarian riders gash themselves so that their king can be honourably lamented in the blood of men, not women's tears, and after Attila is entombed the slaves who do the work are sacrificed to accompany him. Those barbarians were Huns, but on the first occasion when the English are mentioned in history, the Roman historian Tacitus comments on their habit of sacrificing victims to their god or goddess Nerthus by drowning; many preserved corpses have been recovered from the peat-bogs of south Jutland to prove that what he said was true. But the Riders do nothing like that. They are not Christians, but they do not seem to be proper pagans either.

As for the Gondorians, they do have much more sense of ceremony than anyone else in the story, with a custom before meals rather like that of saying grace – they turn and face the West, in memory not only of Númenor but 'beyond to Elvenhome that is, and to that which is beyond Elvenhome and ever will be'. Frodo and Sam characteristically do not understand what Faramir is talking about, nor is it explained, but the Gondorians do believe in the Valar, supernatural powers above the human but below the divine. Tolkien might have had some difficulty in explaining to his Jesuit friend quite what the status of these 'demigods' was, and how they were to be distinguished from pagan deities, and they do lead him into a kind of anachronism, perhaps conscious. When Denethor decides to commit suicide (something of course especially repugnant to Catholics), Gandalf rebukes him:

'Authority is not given to you, Steward of Gondor, to order the hour of your death ... And only the heathen kings, under the domination of the Dark Power, did thus, slaying themselves in pride and despair, murdering their kin to ease their own death.'

(V/7)

Gandalf here faces the idea of funeral sacrifice, and admits that things like this have taken place in Middle-earth, but he uses the significant and in a way illogical adjective 'heathen'. 'Heathen', paradoxically, is a specifically Christian word, the Old English translation of Latin *paganus*, one from the *pagus* or *pays*, not however a 'peasant' in the social sense but a rustic, someone from the back-country, someone ignorant of civilized behaviour, a non-Christian. Does Gandalf calling someone else a heathen imply that he himself is not one, and if so, what is he? The question as usual is not answered.

The deepest sense of religious belief mentioned explicitly in Middle-earth comes in an Appendix, in the 'Part of the Tale of Aragorn and Arwen' told in Appendix A (V). Here we have several death-scenes including that of Aragorn's mother Gilraen, but they are noticeably lacking in what used to be called 'the comforts of religion'. Gilraen dies declaring that she cannot face the 'darkness of our time'. Her son tells her that 'there may be a light beyond the darkness', but she replies only with the riddling *linnod* in Quenya, 'I gave Hope [i.e. Aragorn] to the Dúnedain, I have kept no hope for myself'. Is Aragorn's 'light beyond the darkness' some hope of salvation, some promise of immortality? Or does he just mean that there is a chance still of winning the war which Gilraen sees coming? His own death-scene looks a little further, for Aragorn, talking to his elvish wife, says:

'In sorrow we must go, but not in despair. Behold! We are not bound for ever to the circles of the world, and beyond them is more than memory. Farewell!'

This does suggest that there is another world beyond Middle-earth, and that the 'more than memory' may include a meeting in something like Heaven. But the veiled promise has no effect on Arwen, who has sacrificed her elvish immortality to marry Aragorn. Perhaps the saddest lines in the work are those of her own death in Lothlórien on Cerin Amroth:

> and there is her green grave, until the world is changed, and all the days of her life are utterly forgotten by men that come after, and elanor and niphredil bloom no more east of the Sea.

To revert from emotion to mere grammar, though, the lines are syntactically ambiguous. Do they mean (a) Arwen will lie there until the world is changed; and now she is utterly forgotten? Or (b) Arwen will lie there until the world is changed, and until she is utterly forgotten? If the second alternative is correct, then the changing of the world has already happened, and the sentence means only that Arwen's grave still rests undisturbed. However, the first alternative is also possible, in which case there is a half-suggestion that though Arwen still lies in her grave *and* is utterly forgotten, nevertheless the world-change has not yet taken place: so there is something the other side of 'until', something in our future as well as hers. Tom Bombadil shares the same belief, one might note, for he tells the barrow-wight to go 'where gates are ever shut, / Till the world is mended'. The world, then, *will* be mended, *will* be changed, and then gates will be open, and (perhaps) the dead will rise. The wisest characters in Middle-earth (Gandalf, Aragorn, Bombadil) have some idea of this future resurrection, life after death, but it is never overtly stated, and it is not shared by Arwen or Gilraen, still less by the hobbits. Théoden, like Thorin in *The Hobbit*, has some sense of ancestor-worship, in which the dead go to their fathers, but this is felt only by the most aristocratic characters: Théoden may mean only that he will be buried alongside his predecessors in the row of barrows by Edoras.

In this whole only-slightly-qualified absence of religion the societies of Middle-earth are unlike any human societies we know about; and in this sense Middle-earth could rightly be called a 'Never-never Land'. More politely, and more Catholically, one could say that Middle-earth is a sort of Limbo, in which the characters, like unbaptized innocents or the pagan philosophers of Dante, are counted as neither heathen nor Christian but something in between. Tolkien was moreover not the only writer to set a story in a similar Limbo, the *Beowulf*-epic making very similar censorships (Beowulf's barrow-funeral is like Théoden's, not like Attila's), and exactly the same slip in its single anachronistic use of the word *hæðen*, 'heathen', to condemn Danish devil-worshippers (a point Tolkien considered in his 1936 lecture, see *Essays*, p. 43). The parallel with *Beowulf* may perhaps indicate Tolkien's underlying problem and underlying intention, and cast some light on this paradox of a 'fundamentally Catholic' work which never once mentions God.

Many people have remarked, and even more have felt, that *The Lord of the Rings* is in some way or other a 'mythic' work. The word 'myth', however, has several meanings. One which is certainly relevant is the idea that the main function of a myth is to resolve contradictions, to act as a mediation between or explanation of things which seem to be incompatible. Thus, to give a couple of clear examples, Christians like other monotheists believe in the existence of a God who is at once omnipotent and benevolent; at the same time, no one, not even Boethius, could fail to remark the existence in the world of undeserved suffering, unpunished evil, unrewarded virtue. The incompatibility is resolved by the myth of Adam and Eve, the Garden of Eden and the Fall of Humanity, which explains that evil is the result of human disobedience, and is allowed to exist in order to create free will, freedom to resist or to succumb to temptation, without which humans would be God's slaves, not his children. The whole theory is explained not only by thousands of Christian commentators but also by Milton in *Paradise Lost*, and again by C.S. Lewis in his 1943 rewriting of *Paradise Lost* in twentieth-century terms as *Perelandra*, or *Voyage to Venus*. Meanwhile, if we accept the Norse mythology as retold by Snorri Sturluson, it seems that his pagan ancestors believed in protecting deities like Thor, but also (unlike Christians) that even these deities were not omnipotent, ultimately mortal. The limits of their powers are explained in the myth of the journey of Thor and Loki to the court of the giant-king, in which Thor does his utmost, but is nevertheless not able to overcome the giant-powers of, in succession, Utgarð-Loki, the Miðgarð Serpent, and in the end the hag Elli, or 'Old Age'. Both these stories, the Garden of Eden and the Journey of Thor, look at a contradiction of belief, and tell a story to explain why this is so.

The contrastive nature of myths like these was, however, for Tolkien almost an everyday concern. Virtually every day of his working life as a Professor of Anglo-Saxon, or of English Language and Literature, he found himself reading, teaching, or referring to works like *Beowulf*, the *Elder Edda*, or Snorri Sturluson's *Prose Edda*, which were in one way or other ambiguous in their Christian status. They were almost always recorded by Christians, like the *Beowulf*-poet or (at least two centuries later) the Icelander Snorri. But in some cases they might well have been composed by pagans, like many of the poems of the *Elder Edda*, or they contained explicitly pagan material, like the *Prose Edda*, or they were about heroes whom even the Christian author knew lived in pagan times, and so must have been pagan, like Beowulf himself. To take the last case first, how were heroes of the latter category to be regarded? One of the clichés of *Beowulf*-scholarship is the decision given by the early English churchman Alcuin, who wrote to the abbot of Lindisfarne monastery late in the eighth century, just before that monastery was destroyed by the Vikings, telling him angrily that he must stop his monks listening to stories about pagan heroes – in particular stories about one 'Ingeldus', clearly the same person as a peripheral character in *Beowulf*, Ingeld prince of the Heathobards. He put his opinion rhetorically in the form of question and answer:

> For what has Ingeld to do with Christ? The house is narrow. It cannot contain them both. The King of Heaven wishes to have no fellowship with lost and pagan so-called kings; for the eternal King reigns in Heaven, the lost pagan laments in Hell.

Is that so? For if that is true, then poems like *Beowulf* ought to have been (probably most of them were) ejected from proper monastic libraries. But that is a decision which would have all but destroyed Tolkien's profession, and his ruling passion, and which he could not possibly accept. Moreover, it is clear enough that (though he never mentions Christ) the *Beowulf*-poet was a Christian as devout as his countryman Alcuin: they just had different opinions about the status of pagans, especially of pagans who (unlike the Vikings about to descend on Lindisfarne) had not rejected the Gospel, indeed had never heard of it, or done Christians any harm. The whole poem *Beowulf*, it could be said, is a mediation between contradictory opinions, with strong similarities to *The Lord of the Rings*. Both works were written by believing Christians, but neither mentions anything specific to their belief; *The Lord of the Rings* goes further in all but eliminating references to any form of religion at all. In each case the deaths of heroes remain highly ambiguous. There is a faint sense of hope or consolation for the future at the death of Aragorn, as at the death of Beowulf, but neither work brings this hope into clear focus, and both death-scenes, while dignified, are shadowed by gloom and uncertainty about the future. *Beowulf* further contains not only obscurely menacing references to 'shadow', or 'the Shadow', but also hints of something very like the Valar – for though it is God who sends the hero Scyld to rescue the Danes, he is actually launched on his voyage across the seas by creatures known only as 'those', who must be doing God's will but who may be superhuman. Both works, finally, resolve the same contradiction by a kind of mediation, alien to 'hard-liners' like Alcuin. Must one necessarily believe that all those who lived before the coming of Christ, or between the coming of Christ and the preaching of the Gospel, are irrevocably damned? Neither Tolkien nor the anonymous Old English poet expresses an opinion about this, or indeed ever mentions the question, but both present a heathen or a pre-Christian world with intense sympathy, and with sympathetic bowdlerization (no slavery, no human sacrifices, no pagan gods). The contradiction mediated in both works is that of the 'virtuous pagan': to be damned for inherited paganism or to be saved for personal virtue? In Middle-earth (and one sees that there are more than one reasons for the name) this is a question which need not and in fact cannot arise.

The myth of Frodo

The connection between what has just been argued and the central story of *The Lord of the Rings* lies in the name, and the character, Frodo. There is one very strange thing about his name, and that is that although he ought to have been 'the famousest of the hobbits', his name is one which is never discussed or mentioned at all in the explanation of Shire-names in Appendix F.

Tolkien deals with these at considerable length. Most of them, he says, have been translated from Westron into English according to sense, though in both Westron and English the wearing-down processes of language have left many people unaware that place-name elements like 'bottle' (or 'bold') once meant 'dwelling', so that such names often sound stranger than they once were. When it comes to first names, Tolkien says, hobbits had two main kinds. In category (a) were 'names that had no meaning at all in their daily language', such as 'Bilbo, Bungo, Polo' etc. Some of these were, by accident, the same as modern English names, e.g. 'Otho, Odo, Drogo'. These names were kept, but they were 'anglicized . . . by altering their endings', since in hobbit-names (as in Old English) -a was masculine, -o and -e were feminine. Bilbo's name, then, was actually Bilba. However, there is also a category (b), since in some families it was the custom to give children 'high-sounding' first names drawn from ancient legend. Tolkien says that he has not retained these but translated them, using such faded legendary names as Meriadoc, Peregrin, Fredegar, which do not sound like, but have the same sort of feel as their hobbit originals.

The question is, what sort of name is Frodo – the one name out of all the prominent hobbit characters in *The Hobbit* and *The Lord of the Rings* which Tolkien does not mention or discuss? Possibly the reason is that Frodo is in a one-member category of his own, category (c). His name looks like one of the meaningless ones, such as Bilbo, in which case it would have been, not Frodo, but Froda. However, Froda is *not* a meaningless name. Just like Meriadoc or Fredegar, it is a name from the heroic literature of the past, though it is one which, significantly, and appropriately to Frodo's character, has been all but entirely forgotten. Froda was the father of the hero Ingeld whose legend the monks of Lindisfarne were censured for listening to; *Beowulf* refers to Ingeld, once, as 'the fortunate son of Froda', and that is all we ever heard about Froda in Old English. In Old Norse, though, the exact equivalent of Froda would be *Fróði*, or Frothi, and this name appears frequently and confusingly, as if later authors were trying to make sense of different and contradictory stories. One thing that is certain, though, is that both Froda and Frothi (by rights they should both have a long vowel, *fróda, fróði*) mean 'the wise one' in Old English and Old Norse; and the most prominent of all the Norse Frothis is indeed famous for his wisdom, above all in turning away from war. According to both Saxo Grammaticus (c. 1200) and Snorri Sturluson (c. 1230), this Frothi was an exact contemporary of Christ. During his reign there were no murders, wars, thefts or robberies, and this Golden Age was known as the *Fróða-frið*, 'the peace of Frothi'. It came to an end because the peace really came from the magic mill of Frothi, which he used to grind out peace and prosperity; but in the end he refused to give the giantesses who turned the mill for him any rest, and they rebelled and ground out instead an army to kill Frothi and take his gold. Their magic mill is still grinding at the bottom of the Maelstrom, says Norwegian folk-legend, but now it grinds out salt, and that is why the sea is salty.

Has this story anything, other than the names, to do with either *Beowulf* or *The Lord of the Rings*? One point that may have struck Tolkien is the total contrast between Froda and Ingeld, father and son. The former is a man of peace, the latter the defining image, in the Northern heroic world, of the man who would not give up the obligations of vengeance no matter what this cost him. There is something sad, ironic, and true about the fact that Ingjaldr remained a popular Norse name for generations, and even the monks of Lindisfarne wanted to hear about him, while the story of his peaceful father was rapidly turned into a parable of futility. Frothi, furthermore, is not only a contemporary of Christ, but also an analogue (of course a *failed* analogue), one who tries without ultimate success to put an end to the cycles of war and vengeance and heroism. He fails both personally, in being killed, and ideologically, in that his son and his people return gleefully to the bad old ways of revenge and hatred, and paganism. For after all the 'peace of Frothi' could just have been an accident, an unrealized reflection of the Coming of Christ, about which the pagans never learned. This composite Froda/Frothi, then, could have been to Tolkien a defining image of the

'virtuous pagan', a glimpse of the sad truth behind heroic illusions, a brief and soon-quenched light shining in the darkness
of heathen ages.

All this seems strongly relevant to Tolkien's Frodo. At the start
he is, one may say without impoliteness, a 'good average' hobbit,
no more aggressive than the rest of them – there has never been
a murder in the Shire – but capable of self-defence. He strikes
at the wight in the barrow, tries to stab the Nazgûl on Weathertop,
stabs the troll in the foot in Moria. He thinks it a pity that Bilbo
did not kill Gollum when he had the chance. After Lothlórien,
though, Frodo's actions are increasingly ones of restraint. He
threatens to stab Gollum at in IV/1, but does not do so, and later
on saves him from the archers at 'The Forbidden Pool', against
Sam's strong inclination to say nothing and let him die. He gives
Sting away in VI/2, keeping an orc-blade but saying 'I do not
think it will be my part to strike any blow again'. A few pages
later he throws even that weapon away, declaring 'I'll bear no
weapon, fair or foul'. By this time Frodo has become almost a
pacifist. In 'The Scouring of the Shire' he speaks up defiantly
several times – till the moment when Pippin draws his sword to
avenge the squint-eyed ruffian's insult. Then, though Merry and
Sam also draw and go forward in support, 'Frodo did not move'.
After that he speaks up in defence even of Lotho, reminds the
others that 'there is to be no slaying of hobbits ... No hobbit
has ever killed another on purpose in the Shire', but then in
effect withdraws, saying nothing at all in reply to Merry's 'I knew
we should have to fight'. At the Battle of Bywater he does not
draw sword, and his main concern is to prevent angry hobbits
from killing their prisoners. Even in this there is a touch of
passivity. Talking to Merry (who disagrees and tells him he cannot
save the Shire just by being 'shocked and sad'), Frodo is capable
of giving an order, 'Keep your tempers and hold your hands'.
But as the Battle of Bywater approaches, and Merry blows his horn
and the bystanders cheer, Frodo seems increasingly sidelined:

'All the same,' said Frodo to all those who stood near, 'I
wish for no killing; not even of the ruffians, unless it must
be done, to prevent them hurting hobbits.'

The last two clauses indicate that Frodo has not gone all the way
to pure pacifism (perhaps inconceivable to a man of Tolkien's
background), but 'All the same' seems to concede that an argument has already been lost; 'to all those who stood near' suggests
that Frodo is no longer very assertive even in his rejection of
force. In the end all he says to Merry is 'Very good . . . You make
the arrangements'. He forbids anyone to kill Saruman, and tries
to rescue even the murderer and cannibal Wormtongue, but the
decisions are taken out of his hands first by Wormtongue and
then by the hobbit archers.

All this has its effect on the way he is perceived in the Shire.
As said above, Sam is 'pained' by the way in which Frodo is
supplanted by the large and 'lordly' hobbits, Pippin and Merry,
and by seeing 'how little honour he had in his own country'. It
is prophets who proverbially have no honour in their own
country, and Frodo functions increasingly as a seer rather than
a hero. Even in other countries the honour he has is of the wrong
sort. One remembers Ioreth telling her cousin in Gondor that
Frodo 'went with only his esquire into the Black Country and
fought with the Dark Lord all by himself, and set fire to his Tower,
if you can believe it. At least that is the tale in the City.' The tale is
wrong, but it is a *heroic* tale, and that is the kind of tale people prefer
to hear, 'The better fortitude' (as Milton wrote) 'Of patience and
heroic martyrdom / Unsung'. One wonders what the minstrel said
in the lay of 'Frodo of the Nine Fingers and the Ring of Doom',
but whatever it was, it was forgotten. The end of Frodo's quest, in
the memory of Middle-earth, is nothing. Bilbo dwindles into 'mad
Baggins', a figure of folklore, the elves and dwarves percolate
through to our world in legends of shape-shifters and sword-
makers, even 'the Dark Tower' is remembered in a fragment from

'poor Tom' in *King Lear*. But of Frodo there remains not a trace: except (and one sees that Tolkien has made even the fragmentary nature of his sources into a part of his story) faint hints of an unlucky, well-meaning, ill-fated king, his reputation eclipsed on the one hand by the fame of his vengeful and conventionally heroic son, on the other by the Coming of the true hero Christ, who made the *Fróða-frið*, the 'peace of Frothi', literally marginal.

What has Ingeld to do with Christ, asked Alcuin, and the answer is, obviously, 'nothing'. But Froda has to do with both, father of one, analogue of the other. He is a hinge, a mediation: and so is Tolkien's Frodo, the middle-most character in all of Middle-earth. It would be quite wrong to suggest that he is a Christ-figure, an allegory of Christ, any more than the Ring is one of nuclear power – the differences, as Tolkien pointed out in the latter case and easily could have in the former, are more obvious than the similarities. Yet he represents something related: perhaps, an image of natural humanity trying to do its best in native decency, trying to find its way from inertia (the Shire) past mere furious heroic dauntlessness (Boromir and the rest) to some limited success, and doing it without the resources of the heroes and the *longaevi*, like Aragorn and Legolas and Gimli. He has to do so furthermore by destroying the Ring, which is merely secular power and ambition, and he does so with no certain faith in rescue (or salvation) from outside, from beyond 'the circles of the world'. In this he is once again a highly contemporary figure, an image suitable for a society which as Tolkien knew perfectly well had largely lost religious faith and had no developed theory to put in its place. Could 'native decency' be enough? As a Christian, Tolkien was bound to say 'no', as a scholar of pagan and near-pagan literature he could not help seeing that there had been virtue, and a wish for something more, even among pagans. The myth, or story, that he created expresses both hope and sadness. It is a mark of its success that it has been appreciated by many who share its author's real beliefs, but by even more who do not.

Timeless poetry and true tradition

One of the differences between applicability and allegory, between myth and legend, must be that myth and applicability are timeless, allegory and legend time-constrained. The difference of course is not an absolute one, and a story can have elements of both at the same time: Saruman, and the Master of Laketown, are both examples of something which one can recognize as having a timeless quality, likely to reappear among human beings in any Age of the world, and which one can readily apply to modern times in particular. This does not mean that they stop having roles in a single, one-moment-in-time story, and it would be unfortunate if they did, for they would fade away to becoming mere labelled abstractions. Fortunately there are, scattered through *The Lord of the Rings*, demonstrations of Tolkien's attitude to individual time and to mythic timelessness. They are often related to a subject not yet discussed with relation to either *The Hobbit* or *The Lord of the Rings*, but of major importance to both, and to Tolkien: Tolkien's poetry.

The poetry of the Shire in particular – plain, simple, straightforward in theme and expression – seems continuously variable. The first time we hear the poem listed in the *Lord of the Rings* Index as 'Old Walking-Song' is when Bilbo leaves Gandalf and Bag End in the first chapter, and Bilbo sings it at the door. It is obviously closely related to that particular situation. Bilbo sings:

> The Road goes ever on and on
> Down from the door where it began.
> Now far ahead the Road has gone,
> And I must follow, if I can,
> Pursuing it with eager feet . . .

Bilbo is here singing about something which he is just about to do; and 'the door where it began' in line 2 is the door he is standing at, the door on which Gandalf put the secret mark for the dwarves many years before, when indeed Bilbo's adventures began. The next time we encounter the poem, though, it is Frodo who sings it, just before the hobbits' first encounter with a Ring-wraith, and there are two significant changes: Frodo does not sing, he speaks, and he does not say 'eager feet', he says 'weary feet'. Whose version is correct? Obviously, either. One could say just that Frodo has adapted Bilbo's song to suit his own less happy and hopeful circumstances, but then it is quite possible that Bilbo did the same thing himself. As soon as Frodo has finished, Pippin says: 'That sounds like a bit of old Bilbo's rhyming ... Or is it one of your imitations? It does not sound altogether encouraging.' Frodo replies, 'I don't know ... It came to me then, as if I was making it up, but I may have heard it long ago.' Actually, we know that Frodo is *not* making most of it up, since we have already heard Bilbo's version. But this does not mean that Bilbo made it up, or not that he made *all* of it up. Three pages after Frodo's adaptation, the hobbits start to sing another song, listed in the Index as just 'A Walking-Song', and this time we are told that 'Bilbo had made the words', but set them to 'a tune as old as the hills'.

It should be noted that the two poems, the 'Old Walking-Song' repeated by Bilbo and Frodo, and the longer 'Walking-Song' sung by Frodo and his companions collectively, are quite easy to tell apart: the first one is an eight-line stanza and has alternating rhymes ababcdcd, the second is in ten-line stanza, divided into six longer and four shorter lines, and is in couplets all the way through. Just the same, the Index has mixed them up, and one can see why. For the 'Old Walking-Song' comes back a third time, to be repeated in Rivendell by Bilbo once more, at the end of VI/6. The context here is one of the many sad scenes in *The Lord of the Rings*, for it is obvious to everyone that Bilbo is dying.

His memory has gone, he keeps on falling asleep, he even asks 'what's become of my ring, Frodo, that you took away?', no longer remembering anything of what has happened. As he talks on, with painful irrelevance, he breaks into a third version of the 'Old Walking-Song' or 'Road' poem, this time radically altered:

> The Road goes ever on and on,
>   Out from the door where it began.
> Now far ahead the Road has gone,
>   Let others follow it who can!
> Let them a journey new begin,
>   But I at last with weary feet
> Will turn towards the lighted inn,
>   My evening-rest and sleep to meet.

When Bilbo talks about 'sleep' and 'the lighted inn' he could mean, perhaps does mean, Rivendell and the sleep he falls into as soon as he finishes the poem (just as the 'door' in his first version could be the door of Bag End). But in the Rivendell scene everyone realizes immediately that there is a symbolic meaning, in which the sleep is death. Sam says cautiously and tactfully that Bilbo cannot have been doing much writing – 'He won't ever write our story now' – and Bilbo wakes up enough to respond and in a sense appoint Frodo his literary executor. Bilbo, then, has adapted the poem just as Frodo did (taking over Frodo's phrase 'weary feet'), and left it still immediately relevant to his own personal circumstances, to what is happening in the room at the time. But the more the poem is adapted, the clearer its symbolic sense becomes, in which the Road is life, always followed, eagerly or wearily, but from which everyone in the end must turn aside, leaving it to others to follow.

The counterpart to Bilbo's leave-taking comes in the last chapter, when Frodo, heading for the Grey Havens and the boat which will take him out of Middle-earth altogether, starts to sing 'the old walking-song, but the words were not the same'. The Index lists this, not unnaturally, under 'Old Walking-Song', i.e. the 'Road' poem we have had three times before, but in fact it is the other one, the one in couplets – it should be indexed as 'A Walking-Song' – though it is true that once again 'the words are not the same'. Both versions have the same lines about there being 'A new road or a secret gate', but where the hobbits setting off sang only:

> And though *we pass them by today*,
> Tomorrow we *may come* this way
> *And take* the hidden paths that run,
> Towards the Moon or to the Sun,

Frodo passing out of Middle-earth sings:

> And though *I oft have passed them by*,
> A day will come at last when I
> *Shall take* the hidden paths that run
> West of the Moon, East of the Sun.

Frodo's version is once more, like Bilbo's Rivendell version of the 'Road' song, entirely relevant to his immediate situation, when he is just about to take the 'hidden path' that leads out of the world, but at the same time the 'new road' and the 'secret gate' have taken on quite a different significance. The hobbits may just have meant (as is suitable for a walking-song) that another day they could take a different route, but Frodo's 'new road' is the 'Lost Straight Road' of Tolkien's own mythology, the road to Elvenhome. Shire-poetry, in short, can be new and old at the same time, highly personal and more-than-personal, subject to continuous change while retaining a recognizable frame. It is not surprising that the indexers of *The Lord of the Rings* got confused between poems and versions. But that, one might say, is mythic timelessness for you in miniature. Myths are what is always available for individuals to make over, and apply to their own circumstances, without ever gaining control or permanent single-meaning possession.

Three Shire-poets: Shakespeare, Milton, and 'Anonymous'

The last clause may account for some of Tolkien's expressed annoyance about his poetic predecessors, especially Shakespeare. In Tolkien's professional lifetime Shakespeare had a status which approached the holy, and it has seemed indefensibly Philistine to many critics that Tolkien should have had the nerve to be dissatisfied with him; but Tolkien usually saw things from a different angle than his literary colleagues, and often expressed only half of his opinion at a time. What Tolkien said, in a letter to W.H. Auden, is that at school he 'disliked cordially' Shakespeare's plays (using the same word that he used about allegory), and remembered especially 'the bitter disappointment and disgust . . . with the shabby use made in Shakespeare [in *Macbeth*] of the coming of "Great Birnam Wood to high Dunsinane hill" '.

On the face of it, this does not seem to be true. If there is one work to which *The Lord of the Rings* is indebted again and again, it is Shakespeare's *Macbeth*. Not only do we find the 'march of the trees' motif entirely reworked in the march of the Ents on Isengard, and on Helm's Deep. The prophecy on which the Lord of the Nazgûl relies – 'No living man may hinder me!' – is much the same as the one which the witches' apparition gives to Macbeth, 'Laugh to scorn / The power of man, for none of woman born / Shall harm Macbeth'. Both Macbeth and the Nazgûl are deceived in much the same way, for Macduff was not born but 'from his mother's womb / Untimely ripped', while the Nazgûl falls not to a 'living man', but to the combined attack of Éowyn, a woman, and Merry, a hobbit. The scenes in which Aragorn uses *athelas* to heal the injured recall the account in *Macbeth* of King Edward the Confessor touching for the King's Evil, and healing through his sacred power of royalty; and there is a kind of rebuke to *Macbeth* in the scene in which Denethor discusses the role of stewards and kings. In *Macbeth* Shakespeare is generally thought to have been flattering James VI of Scotland and I of England, who succeeded Elizabeth on the throne in 1603. He claimed descent from Banquo, and some say that in the scene in which the witches show Macbeth Banquo's long line of descendants, the original production in front of the king had a mirror on stage set so as to have King James appear in it. James was, however, a Stewart, or Steward, like Denethor: except that in Scotland, and England, stewards could aspire to the throne. When Denethor replies to Boromir's dissatisfied question – how many years does it take for a steward to become worthy of a vacant throne – that this happens only 'in places of less royalty' (IV/5), the remark applies to Britain. Tolkien could be seen, here as in the march of the Ents, to be correcting or improving on *Macbeth*. He may have had a low opinion of Shakespeare's dramatic opportunism.

The most suggestive contrast with *Macbeth*, however, lies perhaps in the use of magic to foretell the future. The central irony in *Macbeth* is that the witches speak true. Everything they and their apparitions say comes about, though increasingly in ways which Macbeth did not expect. He *is* made Thane of Cawdor; he *does* become 'king hereafter'; the advice to 'beware Macduff' is sound; 'none of woman born' ever does harm him; he is not vanquished till Birnam Wood comes to Dunsinane; and it is Banquo's descendants who succeed, not his own. The question which the play does not raise is whether these events would have come true if Macbeth had not tried to make them come true. He was made Thane of Cawdor without doing anything. Could he have succeeded by fair means without murdering Duncan? And would Macduff have become his deadly enemy if Macbeth had not tried to forestall the prophecy by murdering him and his family? Shakespeare does not raise these issues, but Tolkien does, in the scene in which Galadriel shows the Fellowship her Mirror. One might note that she does not entirely accept Sam's repeated use of the word 'magic', saying that she does not 'understand clearly' what the word means, and that the same word is used to describe 'the deceits of the Enemy'; so the 'magic of Galadriel' is not the same as the 'deceits' of Macbeth's witches. She also adds that 'the Mirror shows many things, and not all have yet come to pass. Some never come to be, unless those that behold the visions turn aside from their path to prevent them.' Someone should have told Macbeth that. But the dilemma is the same in both works. If Macbeth had ignored the witches' deceits, and had refused to murder Duncan, would their prophecy have come true anyway? If not, then they have no power. But maybe it would have, in some quite unexpected way. Similarly, if the Nazgûl had not been faced by Éowyn and Merry – if for instance he had been confronted by Gandalf, as might have happened if Gandalf had not been held back by Pippin, see p. 173 above – would the prophecy about his fate have been falsified? Perhaps not, for it could again have come true in some other way: Gandalf is arguably not a man either, for one thing. Both Tolkien and Shakespeare are aware of prophetic ambiguity, but Tolkien is much more concerned with drawing out its philosophical implications. His point, always, is that his characters have free will but no clear guidance, not from the *palantír*, or from the Mirror of Galadriel. As it happens, all the visions seen in the Mirror by Sam and Frodo seem to be true, though they are a mix of present, past and future; but unlike the witches' visions, they have no effect on anyone's actions.

Tolkien's complex attitude to Shakespeare may now be somewhat clearer. Tolkien, I think, was guardedly respectful of Shakespeare, and may even have felt (if the thought does not seem too sacrilegious to Bardolaters) a sort of fellow-feeling with him. After all, Shakespeare was a close countryman, from Warwickshire, in which county Tolkien spent the happiest years of his childhood, and which in early drafts of his *Lost Tales* mythology he had tried to identify with Elfland. Shakespeare could write Shire-poetry too. Bilbo's poem in 'The Ring Goes South',

> When winter first begins to bite
>   and stones crack in the frosty night,
> when pools are black and trees are bare,
>   'tis evil in the Wild to fare,

is a clear rewrite of Shakespeare's stanzas at the end of *Love's Labour's Lost*:

> When icicles hang by the wall,
>   And Dick the shepherd blows his nail,
> And Tom bears logs into the hall,
>   And milk comes frozen home in pail,
> When blood is nipped, and ways be foul
> Then nightly sings the staring owl . . .

And just as we cannot be sure that Bilbo's poem is his (it might be another Shire-saying 'as old as the hills'), so Shakespeare's stanzas have about them an air, and a by no means unattractive air, of folk-tradition. But the trouble with Shakespeare (Tolkien might have said) was that he was too much of a dramatist. He dealt by choice with single events closely related to the fortunes of particular characters, tightly contextualized. His witches' visions apply only to Macbeth; we are offered no alternative within the text to Macduff as fulfiller of the 'none of woman born' prophecy; the march of the trees is only a tactical ruse – and if one looks at it this way, it is indeed a 'bitter disappointment' that what the Messenger says, 'anon methought / The wood began to move', is no more than a mistake. What Shakespeare did not try to reach in such scenes is the simultaneous immediate relevance, and wider symbolic application, so carefully set up by Tolkien, especially through the device of inset poems. Shakespeare could have done it, of course, and (Tolkien might also have said) he showed his abilities in scenes and characters which Tolkien clearly noted, like the enchanted wood in *A Midsummer Night's Dream* (a model of sorts for Fangorn), or the enchanter Prospero in *The Tempest* (a model of sorts for Gandalf, at least in his short temper). But Shakespeare left the Mark and went to London to seek and find his fortune. His uses of 'true tradition', the traditions of the Shire and the Mark, are accordingly peripheral.

There is a better example of 'mythic timelessness', again linked to native poetic tradition, in the account of Lothlórien. Just before they look in the Mirror of Galadriel, Sam has summed up the peculiar feel of Lórien, saying in effect that it is indefinable. The elves seem to be even more at home there than hobbits in the Shire:

'Whether they've made the land, or the land's made them, it's hard to say, if you take my meaning . . . If there's any magic about, it's right down deep, where I can't lay my hands on it, in a manner of speaking.'

Frodo agrees with him, saying in reply to the last remark, 'You can see and feel it everywhere'. But where does the 'magic', if that is the right word for it, come from? Part of the answer is that it comes from another of the great poets of the Mark, one whom Tolkien rated perhaps higher than Shakespeare, though we do not know his name. The poem *Sir Gawain and the Green Knight*, which Tolkien edited together with his colleague E.V. Gordon in 1925, is found in only one manuscript, and that manuscript contains besides *Sir Gawain* three other poems, almost certainly by the same poet. One of these is called *Pearl*, and Tolkien remained involved with it all his life. The poem is in an extremely complex stanza-form (like many early poems from the West Midlands), which Tolkien carefully and painstakingly imitated in an early poem called 'The Nameless Land', published in 1927. After the edition of *Sir Gawain* came out, the plan was for Tolkien and Gordon to collaborate on an edition of *Pearl*. But Gordon died prematurely in 1938, and the project was taken over by his widow Ida Gordon, who eventually completed and published it in 1953. She thanks Tolkien for his assistance in the 'Preface', and some of the notes to it probably came originally from him, or from his suggestions. Tolkien however kept working on the poem by himself. His translation of it, in the original stanza-form, was published two years after he died, in 1975. One may wonder what was the source of the continuing interest; and what this has to do with myth, and with Lórien.

Pearl seems to be (the whole narration is veiled and riddling) an elegy for a dead infant daughter, possibly called Margaret, which means 'pearl', written by her father. At the start of the poem he goes into an 'arbour' to look for the pearl he has lost there, and falls asleep with his head on a mound. The mound is the child's grave, the arbour the graveyard. In his sleep he finds himself in a strange land where his grief disappears, and where he sees his pearl on the other side of a river. They have a conversation in which she explains the nature of salvation to him, and in the end he tries to rush across the river – only to wake and find himself back in the graveyard, still sad, but now enlightened. All readers realize that the river which the dreamer cannot cross is the river of death. But in that case, where is he standing? What is the strange land, the 'nameless land', with its brilliant trees and shining gravel? It is not Paradise, for that is the *other* side of the river; but it is not Middle-earth either, for in it all grief is forgotten. Already one can see the hints of Lórien, affected by the medieval legends (which both Tolkien and the *Pearl*-poet knew) of the Earthly Paradise.

But the old poem has provided Tolkien with further suggestions. The whole approach to Lórien is an oddly complex one.

First the Fellowship, coming down from the Dimrill Dale, meets the source of the Silverlode, which Gimli immediately cautions the others not to drink. Then they meet the Nimrodel, whose 'falling water', says Legolas, 'may bring us sleep and forgetfulness of grief'. As they cross it, Frodo feels that 'the stain of travel and all weariness was washed from his limbs'. This could mean, of course, entirely literally, that Frodo feels the grime of Moria being washed off, but 'stain' is a slightly odd word to use in this context. It is also an odd word etymologically, the *OED* suggesting that it is originally French, but affected by a Norse word which merely sounded similar. As a result it has the early meaning, not only of 'to colour' or 'discolour', but also almost the opposite, 'to lose lustre'. It is repeated some pages later, when Frodo reaches Cerin Amroth, in a passage of description which seems designed to elaborate exactly the last meaning given:

> [Frodo] saw no colour but those he knew, gold and white and blue and green, but they were fresh and poignant, as if he had at that moment first perceived them and made for them names new and wonderful. In winter here no heart could mourn for summer or for spring. No blemish or sickness or deformity could be seen in anything that grew upon the earth. On the land of Lórien there was no stain.

Gandalf uses the word again in his 'Song of Lórien' much later, 'Unmarred, unstained is leaf and land / in Dwimordene, in Lórien'. So, the 'stain' of normal life is washed off by crossing the Nimrodel; life on the other side regains its natural lustre.

But then the Fellowship goes on and crosses a second river, the Silverlode (the one which Gimli had warned them not to drink). Nor do they wade it this time, they cross above it on the rope-bridge. The reason given by Haldir, again an entirely literal and sensible one, is that it is both 'swift and deep, and . . . very cold'; and Sam, with his fear of heights and vague chatter about ropewalks and his uncle Andy, is there to prevent any immediate hints of allegory, or any more-than-literal meaning. Just the same, there are continuing hints that the rivers which the Fellowship keeps crossing are leading them further and further out of the world. Once they are across the Nimrodel, they are in something like the Earthly Paradise, the place where the *Pearl*-dreamer forgot even the grief for his dead daughter; the members of the Fellowship seem likewise to forget about Gandalf till Celeborn asks them directly. But where are they once they are across the Silverlode, a stream across which Gollum is unable to follow them? One answer is that they are as if dead: at the end of the chapter 'Lothlórien' it says that Aragorn never returned to Cerin Amroth 'as living man'. He did, then, as a dead one? To visit his wife Arwen's grave? Or are they in England – old England, of course, real England, the 'mountains green' of 'ancient time' before the 'dark satanic mills' of Blake's poem? Haldir says very carefully that 'You have entered the Naith of Lórien, or the Gore, as you would say'. We would say neither 'naith' nor 'gore', but Haldir tries a third word with similar meaning when he says they can walk free till they come nearer the heart of the kingdom, 'in the Angle between the waters'. The names 'England' and 'English' come from the word 'angle', and the old now-German homeland of the English was the Angle, or corner of land, between the Flensburg Fjord and the River Schlei – just as that of the hobbits was the Angle between the rivers Hoarwell and Loudwater. Frodo feels that he is 'walking in a world that was no more', that he has 'stepped over a bridge of time'. And perhaps, like the dreamer in *Pearl*, he has.

Tolkien thought that the *Pearl*-poet came from Lancashire, but would be pleased, I think, to hear later arguments that he came from Staffordshire; for Tolkien said repeatedly that he was 'a West-midlander by blood', that he 'took to early West-midland Middle English as a known tongue as soon as [he] set eyes on it' (*Letters*, p. 213), and the heart of the West Midlands is formed by the five counties of Herefordshire and Shropshire, Worcester, Warwick, and Stafford. Like Shakespeare, the Warwickshire man, the Staffordshire *Pearl*-poet was in touch with true traditions, of English poetry, of otherworld vision, of real-world insight, and unlike Shakespeare he had never been lured away from them. His dreamer's state of liminal uncertainty, in which he is both aware of the physical and literal world, and conscious of some deeper symbolic meaning, is exactly the state of mythic or magic timelessness which I believe Tolkien aimed from time to time to reach. And there is one more poet, or poem, which I think contributed to Tolkien's mix of myth and poetry, and that is the rather unlikely figure of John Milton, the Protestant, the regicide, author of the masque *Comus*.

One should note, to begin with, that there is an elvish element in Tolkien's poetry. It is considered in its pure form in the article by Patrick Wynne and Carl Hostetter in *Tolkien's 'Legendarium'*, but it is present even within the poetry of the Shire, which Tolkien explains within the story as the result of Bilbo's contacts with the elves and antiquarian researches. Thus, almost immediately after the hobbits sing the second 'walking song', and then fall silent as they see a Black Rider tracking them, the Nazgûl is driven off by the appearance of a party of elves. They are singing, in Quenya (the older of the two elvish languages used in *The Lord of the Rings*), and Frodo is the only one who understands any of it, consciously. However, 'the sound blending with the melody seemed to shape itself in [all of] their thought into words which they only partly understood'. We then have four stanzas of the song as understood by Frodo, an invocation to Elbereth. The song returns as seven lines of Sindarin (the language of the elves who remained in Middle-earth) sung in Rivendell, though this time the lines are not translated: Frodo merely stands and listens while 'the sweet syllables of the elvish song fell like clear jewels of blended word and melody'. 'It is a song to Elbereth', explains Bilbo. Tolkien is here carrying out a rather daring exercise on

his readers' patience, first by not translating the Sindarin song
and second, by explaining nothing in either case about the subject
of the song, Elbereth. His belief seems to be that, for his readers
as for Frodo's hobbit companions, the sound of the poetry on
its own will convey (some) meaning. In the last chapter of the
whole work, just after Frodo has sung the second altered version
of the second walking song, the elves reply with four lines of
the Sindarin song of Rivendell, this time translated: 'We still
remember, we who dwell / In this far land beneath the trees /
The starlight on the Western Seas.' Meanwhile, as he turns to
face Shelob at the start of the last chapter of *The Two Towers*,
Sam remembers both the elvish song he heard in the Shire, and
the one he heard in Rivendell, so that 'his tongue was loosed and
his voice cried in a language which he did not know' – and he
comes out with yet a third invocation to Elbereth, this time
in Sindarin, but again not translated. Finally, once one has the
translations (Tolkien eventually gave them in 1968 when he contributed the texts to Donald Swann's song-cycle *The Road Goes
Ever On*), one can see that the four Elbereth poems all have a
bearing on two further hobbit-poems, the song Frodo sings to
try to encourage his fellows in 'The Old Forest', and the song
which Sam sings to try to locate Frodo in 'The Tower of Cirith
Ungol', a song the words of which come 'unbidden' to him to
fit a 'simple tune' he knows already – just like the first hobbit
walking song, with its (possibly) new words and tune which is
'as old as the hills'. Sam indeed has just been murmuring 'old
childish tunes out of the Shire', as well as 'snatches of Mr Bilbo's
rhymes', so that one could believe that his 'Song in the Orc-
tower', as the index calls it, is like other Shire-poems part his,
part Bilbo's, and part traditional.

What these six poems have in common (the four elvish Elbereth songs, Frodo's song in the Old Forest and Sam's in the Tower)
is the reflection of a myth. It is a myth in two senses, first, an
old story about semi-divine creatures (Elbereth), though to the

long-lived elves this is a matter of memory and nostalgia rather than mere tradition and belief; and second, with more reference both to the hobbits and the readers, a set of images presenting a world-view. The images oppose stars and trees: the stars give a promise, or for the elves a memory, of a world elsewhere; the trees represent both this world and a barrier to starlight, something through the branches of which mortals look up to try to catch a glimpse of the vision which would otherwise be clear. So the elves address Elbereth as, 'O Light to us that wander here / Amid the world of woven trees', and sing that 'We still remember, we who dwell / In this far land beneath the trees, / Thy starlight on the Western Seas'. The Sindarin song of Rivendell again addresses Elbereth as kindler of the stars, and presents the singer as gazing at the stars *o galadhremmin ennorath*, 'from tree-tangled Middle-earth'. (There is a further confusion here. In *The Road Goes Ever On*, Tolkien gives word-for-word translations between the lines first of the Sindarin song of Rivendell, and then of Sam's cry in Shelob's lair. At the foot of the page he gives connected translations of both. Through some error, though, *o galadhremmin ennorath* is omitted from the latter. I have substituted 'Middle-earth' for the literal 'middle-lands'.)

Sam meanwhile calls to Elbereth *dinguruthos*, 'beneath the horror of death' – and of course this is very relevant to the immediate context, facing Shelob. However, Middle-earth is the world of mortality. The tangle of trees is also a horror. Indeed (to revert to *Comus* for a moment), that is exactly what Milton calls it, 'beneath the horror of this shady wood'.

The horror of the trees is also a matter of immediate context in Frodo's cut-off song in the Old Forest, which begins, 'O! Wanderers in the shadowed land', and goes on to claim that all will emerge in the end from the dark woods and see the sun, 'For east or west all woods must fail . . .' The song is cut off when a branch drops near them, and Merry remarks that the woods 'do not like all that about ending and failing': better to get into the open before asserting the truth of myth. However despite the immediate context, what Frodo also means is something, as usual in Shire-poetry, more general and more symbolic: that the world is like a wood, in which one can easily wander lost and confused, like Aragorn and his companions in the enchanted wood of Fangorn; but that in the end (and in this context that perhaps means after life in Middle-earth is over) all will become clear, as one escapes from both 'the horror of this shady wood' (Milton) and *nguruthos*, 'death-horror' (Sam). Sam gives the obverse of the thought in his Cirith Ungol song, which seems to be sung by a prisoner, like Frodo, 'in darkness buried deep', who nevertheless remembers, like the elves and the hobbits in the forest, that 'above all shadows rides the Sun, / and stars for ever dwell'. Sam's song ends, 'I will not say the Day is done, / nor bid the Stars farewell'. One can say several things about these last two lines. They have of course a point in immediate context, which is to encourage Frodo in his prison not to lose hope. They also repeat the elvish myth of the stars. They further repeat, but strongly contradict, a famous Shakespeare passage from *Antony and Cleopatra*, in which Cleopatra says to her handmaiden, 'Our bright day is done, / And we are for the dark'. But was the phrase 'day is done' Shakespeare's? Certainly not, it must be 'as old as the hills', like the alliterative opposition of 'day' and 'dark' (Old English *dæg, deorc*), free for any true poet writing English to use in any age.

In a similar way Tolkien's whole myth of the stars and the wood is present in embryo in Milton's *Comus*, which tells the story of a maiden lost in a dark wood, and caught by a wicked enchanter who places her in a magic chair, but can do no more because of the preserving powers of her chastity. She is rescued by her brothers, who break in with the aid of a river-nymph (like Bombadil's wife Goldberry) and a protecting plant. But before they meet their supernatural assistant, the two brothers show signs of losing heart. The elder brother prays for moonlight, or for any kind of light that may pierce the 'double night of darkness,

203

and of shades' in which they are wandering. Or if they cannot have light, adds the younger brother, it would be a consolation to hear something from outside the depths of the wood, to remind them there is a world outside:

> ' 'Twould be some solace yet, some little cheering
> In this close dungeon of innumerous boughs.'

'This close dungeon of innumerous boughs' unites at once the images of the elves' *galadhremmin ennorath*, 'tree-tangled Middle-earth', of the hobbits lost in the Old Forest, of Frodo imprisoned in the Orc-tower. As said above, I do not expect that Tolkien had much love for Milton, with his determinedly Protestant epic *Paradise Lost* and his revolutionary political views, but he accepted him like Shakespeare as a poet capable of true poetry; and while Milton was no West Midlander, *Comus* was written for a patron who came from Ludlow in Shropshire, another of the West Midlands core counties, and first performed there: maybe some of the ambience had rubbed off.

Just to round out the connections, there is a mysterious line in *Pearl* where the dreamer says the jewels in the stream of the strange country are bright 'As stremande sterneȝ quen stroþe-men slepe', 'as streaming stars when *stroth*-men sleep'. But what in the world (or out of it) are 'stroth-men'? The note in Ida Gordon's edition (and I believe that some of these came originally from Tolkien's suggestions) explains Old English **stroð* as 'marshy land (overgrown with brushwood)', but explains further that 'stroth-men' must mean, generally, 'men of this world', unaware of any higher one, but also carries pictorially 'a suggestion of the dark, low earth on to which the stars look down'.

The *Pearl*-poet, then, from Staffordshire, saw the inhabitants of Middle-earth as men sleeping in the wood, ignoring the stars above their heads; Milton, writing for Shropshire, produced what is close to an allegory of life as a march through the trees to rescue the imperilled soul; Shakespeare, from Warwickshire, produced his enchanted wood in *A Midsummer Night's Dream*, and knew more of true tradition than he admitted. Tolkien, who saw his family roots as being in Worcestershire, must have felt that he was only voicing, or disentangling, a myth long latent in the poetry of the Mark, and putting it into both the simple poetry of the Shire and the more complex elvish poetry that underlies it, as the lost tradition of poems like *Pearl* underlies unnoticed much of the English poetry that has succeeded it. The essence of this myth, however, is that it always has an immediate point within the context of what is happening at that moment in *The Lord of the Rings*, but carries strong suggestion of far more general, indeed universal, applicability outside that context. Tolkien's myth of stars and trees presents life as a confusion in which we all too easily lose our bearings and forget that there is a world outside our immediate surroundings. This is not incompatible with Christian belief, but perhaps rings more true for those who have no access to Revelation, like the inhabitants of Middle-earth, or who have for the most part forgotten it, like the inhabitants of contemporary England, in Tolkien's lifetime and still more in my own.

One might add finally that it contains a further element of ambiguity. The trees and the forest in this myth are symbols of error, or horror, or death, or confusion. But there are few people who have loved trees more than Tolkien did. The hobbits' second walking song envisages leaving the world for the 'hidden paths that run / Towards the Moon or to the Sun', but the price of this is bidding farewell to the trees, the little and friendly trees of English orchards and hedgerows, 'Apple, thorn, and nut and sloe, / Let them go! Let them go!' Frodo is taken away in the end to Elvenhome to be cured. But he also loses any hope of revisiting Lothlórien, the Earthly Paradise. And as Haldir says as they walk into Lothlórien and he considers the possibility of leaving Middle-earth, there may be a refuge for the elves elsewhere, 'But if there are mallorn-trees beyond the great Sea, none have reported it'. Unlike the Christian myth, Tolkien's myth contains a deep love for and attachment to the beauty of Middle-earth itself, expressed also by Fangorn's sad song of Ent and Ent-wife, and Bregalad's lament for the rowan-trees. The forest, and Middle-earth, can turn into Mirkwood, 'where the trees strive one against another and their branches rot and wither', or into Lórien, so beautiful that in it no grief has power. The image of our world as it is now is perhaps Ithilien, once 'the garden of Gondor' but now part desolate, yet still retaining 'a dishevelled dryad loveliness'; though protesters who write 'Another piece of Mordor' on yet one more example of ruinous chain-saw development have taken one of Tolkien's points. Middle-earth is intrinsically beautiful, and that makes it hard to leave, even for a believer like Tolkien.

Moments of eucatastrophe

There is however one moment in *The Lord of the Rings* where the Christian myth comes close to the surface and is explicitly alluded to; though it may confirm rather than deny the argument of this chapter that it is a moment which almost no one notices, and which looks designed not to be noticed. In his essay 'On Fairy-Stories', reprinted in *Essays*, Tolkien introduced the notion of (his own coinage) 'eucatastrophe', which he defines as 'the good catastrophe, the sudden and joyous turn' – not an ending, for 'there is no true end to any fairy-tale', but the moment when, in Andrew Lang's late, literary and unsatisfactory tale 'Prince Prigio', the dead knights come back to life, or when, in the Scottish tale of 'The Black Bull of Norroway', the heroine's final appeal to her enchanted lover, 'And wilt thou not wauken and turn to me', is answered, 'He heard and turned to her'. In moments like these, Tolkien wrote, we get a glimpse of joy 'that for a moment passes outside the frame, rends indeed the very web of story, and lets a gleam come through': a gleam of revelation from outside the narrative.

In *The Lord of the Rings* this eucatastrophic moment comes in the chapter 'The Field of Cormallen'. In the preceding chapter Frodo and Sam (and Gollum) destroyed the Ring, and at the start of the 'Cormallen' chapter the army of the West feels the realm of Sauron crumble, and sees the Sauron-shape stretching out 'a vast threatening hand' towards them, 'terrible but impotent', for as with Saruman's wraith later on, 'a great wind took it, and it was all blown away, and passed'. Gandalf calls the eagles to take him to Mount Doom, but meanwhile we are returned to Frodo and Sam, who know nothing of what is happening outside. They have little or no hope of getting away (Sam has little, Frodo has none), and in the end they fall unconscious, 'worn out, or choked with fumes and heat, or stricken down by despair at last, hiding their eyes from death'. In their sleep they are picked up and carried away by Gandalf and the eagles. And when they recover consciousness, two weeks later, Sam naturally has no idea what has happened. He seems to be back in Ithilien. But the first person he sees is Gandalf, whom he last saw being dragged into the chasm of Moria by the Balrog, and whom he has long assumed to be dead. Is Gandalf dead? Is Sam dead? Perhaps he has died and gone to heaven (if one could use such a term in Middle-earth). Or has heaven turned Middle-earth into the Earthly Paradise? Sam is like the dreamer-father at the start of the vision in *Pearl*: he does not know where he is. And it is significant that we are given this moment from the viewpoint of Sam, not Frodo (who had woken up before), because Sam's bewilderment is the greater and the more innocent. What he says is: 'Gandalf! I thought you were dead! But then I thought I was dead myself. Is everything sad going to come untrue? What's happened to the world?'

Sam is not dead, nor is everything sad 'going to come untrue'. What Gandalf replies is that 'A great Shadow has departed' – but it is not *the* great Shadow. Gandalf however goes on very carefully to tell Sam what day it is:

> The fourteenth of the New Year ... or if you like the eighth day of April in the Shire reckoning. But in Gondor the New Year will always now begin upon the twenty-fifth of March when Sauron fell, and when you were brought out of the fire to the King.

No one any longer celebrates the twenty-fifth of March, and Tolkien's point is accordingly missed, as I think he intended. He inserted it only as a kind of signature, a personal mark of piety. However, as he knew perfectly well, in old English tradition, 25th March is the date of the Crucifixion, of the first Good Friday. As Good Friday is celebrated on a different day each year, Easter being a mobile date defined by the phases of the moon, the connection has been lost, except for one thing. In Gondor the New Year will always begin on 25th March, and the same is true of England, in a sadly altered and declined fashion. When the Julian calendar gave way to the Gregorian in 1752, there was an eleven-day discrepancy between them, so that the 25th March jumped to being the 6th of April. And in England the year still *does* start on the 6th of April. But only the tax year, which no one sees as a moment of eucatastrophe.

25th March remains a date deeply embedded in the Christian calendar. In old tradition, again, it is the date of the Annunciation and the conception of Christ – naturally, nine months exactly before Christmas, 25th December. It is also the date of the Fall of Adam and Eve, the *felix culpa* whose disastrous effects the Annunciation and the Crucifixion were to annul or repair. One might note that in the Calendar of dates which Tolkien so carefully wrote out in Appendix B, December 25th is the day on which the Fellowship sets out from Rivendell. The main action of *The Lord of the Rings* takes place, then, in the mythic space between Christmas, Christ's birth, and the crucifixion, Christ's death. Is this telling us something about Frodo? Are we meant to see him as a 'type' of Christ? I do not think so. If Frodo is a 'type', he is so only in a technical sense which has been almost entirely forgotten, and in which the differences are more important than any similarities. Frodo offers no promise of soul-salvation (though he has saved Middle-earth from a great danger), he releases no prisoners from Hell (though he does from Sauron's dungeons), he does not rise from the dead (though Sam for a moment, and entirely understandably, thinks something like that might have happened). Frodo in other words has no supernatural dimension at all. But he and Sam do have a 'eucatastrophic' one.

Tolkien continued the eucatastrophe with the description of the feast of Cormallen, and gave another view of it in the next chapter, when Éowyn and Faramir, left behind in Minas Tirith, also feel the crash of the fall of the Dark Tower. They naturally do not know what it means, and feel it as 'the stroke of doom'. To Faramir it brings thoughts of the Fall of Númenor, 'the great dark wave climbing over the green lands ... darkness unescapable', but he rejects the thought. And then the eagle comes and announces what has happened in a strange verse which is composed, uniquely for Middle-earth, in exactly the language of the Psalms in the Authorized of King James version of the Bible, instantly recognizable to anyone of Tolkien's generation:

Sing and rejoice, ye people of the Tower of Guard,
for your watch hath not been in vain,
and the Black Gate is broken,
and your King hath passed through,
    and he is victorious.

This is in one way rather like the poetry of the Shire. It has immediate contextual meaning. The 'people of the Tower of

Guard' are the garrison of Minas Tirith; 'the Black Gate' is the Morannon, the northern entry to Mordor; 'your King' is Aragorn. At the same time there are strong hints of universal meaning. The image of people guarding a city is commonly applied, in familiar hymns like Martin Luther's *Ein' feste Burg ist unser Gott*, 'A safe stronghold our God is still', to Christians guarding the city of salvation; Christians are often urged in parable to 'watch', to stay awake, never knowing when the Second Coming will take place; 'the Black Gate is broken' could be applied to the Harrowing of Hell, which took place between Good Friday and Easter Monday, between death and resurrection, when Christ led the souls of the patriarchs and prophets out of infernal bondage. Of course, and again, the eagle *does not say that*, and what he does say is adequately explained (just like Frodo singing about escape from the forest in the Old Forest) by what is going on in the immediate story. But both with the eagle and with Frodo, the hints of something greater do not go away: they promote the sense of mythic timelessness.

The Lord of the Rings, then, contains within it hints of the Christian message, but refuses just to repeat it. The myths of Middle-earth furthermore determinedly reject any sense of ultimate salvation. The 'myth of stars and trees' is highly ambiguous about ever escaping from the 'tree-tangle', in part because the inhabitants of Middle-earth do not want to, they want to live on in the woods of the Shire or the forests of Fangorn or Lórien or in the valley of the Withywindle. This hope is clearly not going to be fulfilled. *The Lord of the Rings* indeed seems to be full of 'alternative endings'. There is one in Frodo and Sam's experience. Though they are rescued by eucatastrophe and the eagles, there is a strongly-realized moment when they think they are dead. 'Well, this is the end, Sam Gamgee', says Frodo, twice repeating the phrase 'the end of all things'. Sam tries to tell him there is hope yet, but Frodo replies – and what he says retains a kind of conviction, even after eucatastrophe, 'it's like things are in the world. Hopes fail. An end comes.' This remains generally true, even if the story this one time falsifies it. As Tolkien said of eucatastrophe in 'On Fairy-Stories', surely in 1947 glossing his own fable (his words, but my emphases added):

> In its fairy-tale – or otherworld – setting, it is a sudden and miraculous grace: *never to be counted on to recur*. It does not deny the existence of *dyscatastrophe*, of sorrow and failure: the possibility of these is necessary to the joy of deliverance; it denies (*in the face of much evidence*, if you will) universal final defeat and in so far is *evangelium*, giving a fleeting glimpse of Joy, *Joy beyond the walls of the world*, poignant as grief.
>
> (*Essays*, p. 153)

It should be added that most of the characters in *The Lord of the Rings* are staring 'universal final defeat' in the face. The Ents are doomed to extinction, and oblivion – their fate is proved by the fact that even the Anglo-Saxons did not know what Ents were, though they remembered the word. According to *The Hobbit*, hobbits still exist, but there is certainly no Shire any more. What happens to the elves? Galadriel is sure that they will 'dwindle', and she may mean that they will physically shrink in size, to become the tiny creatures of *A Midsummer Night's Dream* and popular imagination. Or they may dwindle in number. Or something else may happen to them. Tolkien knew the Rollright Stones, the stone circle on the border of Oxfordshire and Warwickshire, and mentions them allusively in *Farmer Giles of Ham*. There is a legend attached to them, which is this. Once upon a time there was an old king, who was challenged by a witch to take seven strides over the hill and look down into the valley beyond. He did, but found his view blocked by a barrow and the witch's curse activated:

Rise up, stick, and stand still, stone,
For king of England thou shalt be none.
Thou and thy men hoar stones shall be,
And I myself an eldern-tree.

This is proper Shire-poetry on several levels. But maybe *that* is what happens to the elves. The last we see of Galadriel and her company (other than the final scene en route to the Grey Havens) is her and Celeborn and Elrond and Gandalf talking after the hobbits are asleep. But do we see them, and are they talking?

If any wanderer had chanced to pass, little would he have seen and heard, and it would have seemed to him only that he saw grey figures, carved in stone, memorials of forgotten things now lost in unpeopled lands. For they did not move or speak with mouth, looking from mind to mind.

The next day the people of Lórien leave, 'Quickly fading into the stones and the shadows'. Fading, or turning? A possible conclusion for the elves is that they do not all leave Middle-earth. Instead, like the old king of Rollright, they are absorbed into the landscape, becoming the 'grey figures, carved in stone', which dot English and Scottish folk-tradition (the Old Man of Coniston, the Grey Man of the Merrick). It would not be an unsuitable, or an entirely sad ending. But it is the marker of an ultimate loss and defeat.

Further mythic moments

The closer the myths of Middle-earth approach to the Christian one, it seems, the sadder (because the more finally inadequate) they become. Tolkien's pre-Christian Limbo contains no real heathens, but it has no scope either for a *Divina Commedia*, a divinely-inspired happy ending. Some of its characters, and not only the failing ones like Denethor but also the victorious ones like Frodo or Fangorn, seem to be on the edge of a situation of existential despair. Yet this is not the impression the work makes as a whole. One of the reasons for its success has certainly been its good humour, its ability to balance loss and defeat with acceptance, optimism, even defiance. I conclude this section by looking at four moments (out of a large possible selection) in which one can see *The Lord of the Rings* carrying out its function as a mediator between, on the one hand, Christian belief and the literature of the pre-Christian heroic world to which Tolkien was so much attached; and on the other, between Christian belief and the post-Christian world in which Tolkien thought himself increasingly to be living.

The first of these is the scene at the gate of Minas Tirith at the end of the chapter, 'The Siege of Gondor'. At this moment several strands of the story are about to come together. Gandalf is waiting at the gate to confront the chief of the Nazgûl, who has just directed against it the battering-ram, Grond. Pippin is running up to get him to come and rescue Faramir. Outside, Merry and the Riders under Théoden are about to arrive, unknown to Gandalf and the defenders. The Lord of the Nazgûl rides in, and is confronted by Gandalf, who tells him to go back, to 'fall into the nothingness that awaits you and your Master'. But the Black Rider takes his challenge and throws back his head, to reveal that 'nothingness' has already come: 'behold! he had a kingly crown, and yet upon no head visible was it set.' He laughs and tells Gandalf, 'Old fool! This is my hour. Do you not know Death when you see it?' (As said above, pp. 129–30, he is at this moment very like Milton's description of Death in *Paradise Lost* Book II.) Gandalf does not reply:

And in that very moment, away behind in some courtyard of the City, a cock crowed. Shrill and clear he crowed, recking nothing of wizardry or war, welcoming only the morning that in the sky far above the shadows of death was coming with the dawn.

And as if in answer there came from far away another note. Horns, horns, horns. In dark Mindolluin's side they dimly echoed. Great horns of the North wildly blowing. Rohan had come at last.

At this moment the Lord of the Nazgûl represents both the Boethian and the Manichaean views of evil, as I have called them, at the same time. Evil does not exist, it is an absence, as Gandalf says, and as the Nazgûl confirms by throwing back his hood. But the absence can have power, can be a force itself, working physically as well as psychologically: this is the essence of the Nazgûl's challenge, to which Gandalf makes (can make?) no answer.

The answer is made instead by the cockcrow, and by the horns. What does the cockcrow stand for? In the Christian story, of course, it is associated with Peter's denial of Christ. Frightened after the arrest of Jesus, Peter denies three times that he knew him, and remembers Christ's prophecy – 'before the cock crows thou shalt deny me thrice' – only after the third denial, when he hears the cock crow and realizes too late what he has done. In that story the crowing of the cock acts above all as a rebuke of Peter's natural fear of death. What it means, perhaps, in that context, is that from now on the fear of death will be conquered, and not only by Peter: for beyond death there will be a resurrection. The Younger Brother in *Comus* imagines cockcrow as something similar. In the dark wood where he and his brother are wandering, he says, it would be a reassurance to hear a cock crow from outside, from beyond the wood:

    ' 'Twould be some solace yet, some little cheering
    In this close dungeon of innumerous boughs.'

Tolkien might well have remembered yet another scene from Northern pagan myth. Saxo Grammaticus tells the story of how King Hadding is guided by a witch to the boundary of the *Ódáins-akr*, 'the Field of the Undying', but he cannot gain entrance. As he turns away the witch beheads a cock and throws it over the boundary. A moment later he hears it crow, alive again. In all the stories the sound means new day, new life, escape from fear and the horror of death.

And in reply, or 'as if' in reply, come the horns blowing. Warhorns are the instrument *par excellence* of the heroic Northern world. In *Beowulf* the nearest thing the poem has to a 'eucatastrophic' moment is the one when the demoralized survivors of Beowulf's nation, the Geats, trapped in Ravens' Wood by Ongentheow, the terrible old king of the Swedes, who has passed the night by shouting threats of what he will do to them in the morning, hear *samod ærdæge*, 'with the dawn', the horns and trumpets of the army of Beowulf's uncle Hygelac coming to their rescue. In later history the men of the Alpine cantons of Switzerland kept horns with special names (like the Nazgûl's battering-ram Grond), the 'Bull' of Schwyz and the 'Cow' of Unterwalden: chronicles tell of them blowing defiantly through the night as the Swiss rallied after disaster at the Battle of Marignano. Roland's horn Olifant is famous, though he is too proud to blow it and call for help. The later chivalric world turned away from them, preferring what *Sir Gawain* calls the 'nwe nakryn noise', the noise of the newly-invented (and Turkish-derived) kettledrums. But in *The Lord of the Rings* Boromir's horn still has the old meaning. Boromir blows his mighty aurochs-horn when he sets out from Rivendell, and Elrond rebukes him for it, to get the defiant answer, 'though thereafter we may walk in the shadows, I will not go forth as a thief in the night'. He blows it again in challenge as the Balrog comes up to the Bridge of Khazad-dûm, and even the 'fiery shadow' checks at it. If the cockcrow means new day, resurrection, and hope, horns mean defiance, recklessness, going on even when there is no hope: two answers to the existential dilemma posed by the Nazgûl, and it may be that the pagan or pre-Christian one is the stronger. It checks the Nazgûl, as it checks the Balrog.

How much of this does the reader need to know? Much of it, like Boromir's horn, and the horn of Gúthlaf in chapter 5, is already in the story and unmistakable. Other images, like the one of dawn coming, are too familiar to need an explanation. The scene can be taken as just a string of coincidences, with the cock crowing because it does, not aware of anything that is going on, and the horns blowing only 'as if' in reply, not connected with the Nazgûl or the problem of 'nothingness' made visible at all. But it is a dull reader here who sees only immediate context.

The same is true of a scene which seems to go the other way, leaning towards despair as the gate-scene does towards defiance: Frodo and Sam in the Dead Marshes. The hobbits are picking their way through these, guided by Gollum, when they start to see the will o' the wisp, the 'misty flames' of marsh-gas. Gollum calls the phenomenon corpse-candles. Sam notices that Frodo seems mesmerized by them, and tells him not to look. Then he too trips and falls with his face in the water, to jump up in horror. 'There are dead things, dead faces in the water'. Frodo, still speaking 'in a dreamlike voice', agrees:

'I have seen them too. In the pools when the candles were lit. They lie in all the pools, pale faces, deep deep under the dark water. I saw them: grim faces and evil, and noble faces and sad. Many faces proud and fair, and weeds in their silver hair. But all foul, all rotting, all dead. A fell light is in them ... I know not who they are.'

Gollum has a simple explanation, 'There was a great battle long ago', the Battle of Dagorlad indeed, and these are the casualties lying in their graves. But Sam does not believe him, 'The Dead can't be really there', and he seems to be right, for Gollum has put his theory to the test and tried to dig down to the graves, without success: 'you cannot reach them. Only shapes to see, perhaps, not to touch'.

What do these faces mean? The ominous thing about them is that they are all, now, the same. They seem to represent the casualties of both sides, the servants of Sauron, 'grim faces and evil', the Elves and Men who opposed and defeated him, 'noble faces and sad'. But it has all come to the same thing in the end. The whole sequence has reminded many people of First World War battlefields (Tolkien was on the Somme for three months), where the static warfare left the dead unburied for years, with both sides inextricably confused. This might account for the fact, the unsurprising fact, that the bodies of both sides corrupt in the same way, they all end up 'rotting' and 'dead'. But in Frodo's vision even the 'noble faces' are 'sad', and they are all not just 'rotting' but 'foul'; they all have a 'fell light' in them. There are several unvoiced implications. That the whole thing has been for nothing (a thought never too far away from the living characters' sense of 'ultimate defeat'); that Sauron, though defeated in battle, has somehow managed to take his revenge on the dead, and now holds them in his grip; perhaps worst of all, that all the dead are hostile to the living, that they have learned something in death that they did not know alive. As said before, there are hints in the barrow-wight chapter that the wight still controls the dead buried in his barrow, may even himself be one of the dead, one of those who fought the Witch-king of Angmar, now turned to evil by some sort of psychic decomposition. A fear like this is powerfully expressed by the Un-man in Lewis's *Perelandra*, the Un-man who is Weston the scientist taken over by diabolic possession: but the awful thing there is that Weston's own psyche seems to be still alive underneath the possession, and screaming for help, terrified that he is going to sink down to what he sees as the inevitable fate of all who die. The conception is a Classical and a heathen one, going back to Homer, and Lewis and Tolkien and all the Inklings no doubt vehemently rejected it. But they did not forget it. Could it be true? Sam in fact suggests that this is 'some devilry hatched in the Dark Land', an illusion, a sending intended to do just what it does, to cause fear and demoralization, and that is the comforting answer. The right thing to do is what the hobbits do, press on regardless. But the stain of the vision remains. The defiance of the horns is one image in *The Lord of the Rings*, but the Dead Marshes provide a memory of all that has to be defied. They are two sides of the same existential situation, in a world which does not yet know salvation, and each is the stronger for the other.

There is an analogue to the dilemma just proposed in a highly understated and underplayed scene in Book V/9, 'The Last Debate'. Legolas and Gimli are walking through Minas Tirith sightseeing. Gimli is critical of the stonework, 'some good . . . and some that is less good'. Legolas is rather more appreciative, and remarks that if Gondor can still produce men like Imrahil, in its decline, then it must have been great indeed in its prime. The good stonework is probably the older, says Gimli, half-agreeing, but then going on to generalize:

'It is ever so with the things that Men begin: there is a frost in Spring, or a blight in Summer, and they fail of their promise.'

'Yet seldom do they fail of their seed,' said Legolas. 'And that will lie in the dust and rot to spring up again in times and places unlooked-for. The deeds of Men will outlast us, Gimli.'

'And yet come to naught in the end but might-have-beens,' said the Dwarf.

'To that the Elves know not the answer,' said Legolas.

By this time they are no longer discussing stonework. There is a strong sense that they are foreseeing also the end of the Third Age, and the future domination of Man. But could it also be that these two proverbially soulless creatures, elf and dwarf, are actually discussing (without, of course, being in the slightest aware of it), the Incarnation, the Coming of *the Son of Man*? There is a strong parable element in Legolas's image of the seed, and what he says about the 'deeds of Men' outlasting his and Gimli's species has come true, in our world. Nevertheless Gimli's pessimistic reply might be seen as true also. It would be entirely true without qualification, in the Christian view, if fallen humanity had not been rescued by a Power from outside, a Power beyond humanity which nevertheless became human. But as Legolas says, the elves know nothing about that. Or as Tolkien put it in his Fairy-Stories essay, 'elves are not primarily concerned with us, nor we with them. Our fates are sundered'. Legolas and Gimli go on to tell the hobbits their story about following the Paths of the Dead; and Gandalf, in the 'last debate' itself, reminds everyone that 'it is not our part to master all the tides of the world'. After the momentary glimpse from 'outside the frame', the characters return to the business, the inevitably limited business, of Middle-earth.

The place where Middle-earth comes closest to twentieth-century life, however, is certainly the Shire, and the instinct which led commentators to see 'The Scouring of the Shire' as in some way a comment on Tolkien's own time and country was not entirely false. Rather than seeing it just as an allegory of England in the aftermath of war, however, one might apply what is said there to a more general situation: of a society suffering not only from political misrule, but from a strange and generalized crisis of confidence. A similar diagnosis was made about England in entirely realistic terms by Tolkien's great contemporary writer of fable, George Orwell, though he did it not in *Nineteen Eighty-Four* but in his relatively neglected between-war novel, *Coming Up for*

Air (1938). In this the odd, the inexplicable thing, is that although the lead character George Bowling knows perfectly well what he wants to do with his life (go fishing), he never gets the chance till too late, and when he does, the idyllic world of childhood he remembers has completely vanished under suburban 'development', pools, fish, town, social life, community, all together. But why did he tamely acquiesce in the frittering away of his life and his hopes? Why, to return to Middle-earth, do the hobbits of the Shire tamely allow themselves to be taken over, when they quite clearly have the strength to resist, and face very little opposition when they do resist? They have no leadership; they are bewildered; they (or some of them) are like the Riders, confused by the Voice of Saruman, the insistent persuasion of modern political jargon. To this the answer, inside *The Lord of the Rings*, is the horn of Eorl the Young, made by the dwarves, taken from the hoard of Scatha the Worm, given to Merry by Éowyn. 'He that blows it at need,' she says, 'shall set fear in the hearts of his enemies and joy in the hearts of his friends, and they shall hear him and come to him.' In the Shire the rebellion starts as soon as Merry blows it, saying, 'I am going to blow the horn of Rohan, and give them all some music they have never heard before.' Immediately the paralysis dissipates. Everyone seems to wake up. Not only do they know what they want (they always did, like Orwell's George Bowling), they have no hesitation in getting it, and rejecting the casual, pointless destruction (pouring filth into the streams, felling all the trees along the Bywater road, cutting down the Party Tree) that comes with Saruman and all he stands for.

Inside *The Lord of the Rings*, the horn of Rohan stands for a rejection of the despair which is Sauron's chief weapon, and which hangs persistently on the edges of the story, in the barrow, in the Dead Marshes, in Fangorn Forest, in Mordor, and even in the Shire. Outside *The Lord of the Rings*, it stands maybe for *The Lord of the Rings*. If Tolkien were to choose a symbol for his story and its message, it would be, I think, the horn of Eorl. He would have liked to blow it in his own country, and disperse the cloud of post-war and post-faith disillusionment, depression, acquiescence, which so strangely (and twice in his lifetime) followed on victory. And perhaps he did.

Style and genre

One final short comment should be made about the genre of *The Lord of the Rings*. In one obvious way it has created its own genre. The heroic fantasy trilogy – a genre, or sub-genre, totally unknown till Tolkien wrote one – has now become a publishing staple, obviously in imitation and emulation of *The Lord of the Rings*. Is it, however, still also a novel? Or a romance? Or even an epic? The difficulty of deciding tells us something about it.

The most comprehensive description we have of literary modes is that of Northrop Frye, in his book *An Anatomy of Criticism*, which came out only just after *The Lord of the Rings*, in 1957. Frye never mentions Tolkien's work in the *Anatomy*. However, the framework he gives both allows us to place *The Lord of the Rings*, and to see why it is an anomaly. In Frye's view, there are five very general literary modes, defined only by the nature of their characters. At the top is *myth*: if the characters in a work are 'superior in kind both to other men and to the environment of other men', Frye declares, then 'the hero is a divine being and the story about him will be a myth'. One level down is *romance*: here the characters are superior only in 'degree' (not 'kind') to other men, and again to their environment. The next level down is *high mimesis*, the level characteristic of tragedy or epic, where the heroes and heroines are 'superior in degree to other men but not to [their] natural environment'. Next to bottom is *low mimesis*, the level of the classical novel of Jane Austen or Henry James, where the characters are very much on a level with ourselves in abilities, though maybe not in social class. Below it comes *irony*, where we see ourselves looking down on people weaker or more ignorant than ourselves, where heroes turn into anti-heroes and are often treated comically.

Where does *The Lord of the Rings* fit on this schema? The obvious answer is, at all five levels. The hobbits, for a start, are very clearly *low mimetic*, at least most of the time. As discussed in chapter I, their constant engagement with characters on higher levels may pull them up towards heroic speech, and dress, and action, but this is often seen by themselves as odd: Gaffer Gamgee looks at his son's armour and says, unimpressed, 'What's come of his weskit? I don't hold with wearing ironmongery'. Sam in particular (even more than Gollum/Sméagol) tends to sink towards the *ironic*, indeed his relationship with Frodo contains a hint of the most famous ironic/romantic pairing in literature, that of Don Quixote and Sancho Panza. With his proverbs, his common-sense (his name is Samwise, Old English *sám-wís*, 'half-wise'), his stubborn unthinking practicality, he tends always to drop the stylistic level of scenes he is in, even scenes tending towards the mythic like the rope-crossing into Lórien.

Nearly all the human characters occupy a higher level. Éomer, for instance, or Boromir, are characteristic figures of *high mimesis*, leaders, kings, stronger and bolder than everyday life, but still mortal, without supernatural powers. Aragorn, however, though staying on their level much of the time, with an element of deliberate disguise, is different: he can summon the dead, he can compel the *palantír* to his will, he lives in full vigour for 210 years, and is able to control his death. He, his non-human companions like Legolas, Gimli, and Arwen, and all the non-human species of Middle-earth, are figures of *romance*. Finally, characters like Gandalf, Bombadil, and Sauron, are very close to the level of *myth*. They are not exactly 'divine beings', but they are not human either, something intermediate (in fact Gandalf and Sauron are both Maiar, a class of being invented by Tolkien). The whole story furthermore aspires in places to mythic meaning, as discussed above. The aspiration is limited only by Tolkien's refusal to reach out to, to do any more than hint at, a sixth level above and outside Frye's categorizations, which one could call 'true myth', or gospel, or revelation, or (Tolkien's word) *evangelium*. In his essay 'On Fairy-Stories' he argued that fairy-tales could have a glimpse or gleam of this, through 'eucatastrophe', but should not allow it in, as rending 'the very web of story'.

In brief, then *The Lord of the Rings* is a romance, but one which is in continuous negotiation with, and which follows many of the conventions of, the traditional bourgeois novel. All the levels however interact continually and challengingly. Gandalf, Aragorn, Théoden, Merry and Pippin, can all be found together in scenes like the arrival at Isengard, representing as it were all five of Frye's levels at once, and they move easily from amiable hobbitic prattle about pipeweed to 'the speech of the oldest of all living things', and intensely suggestive remarks about the nature of 'chance' and 'luck'. Théoden himself is hard to put on the Frye scale. On the face of it he is the same as his nephew Éomer, but he also becomes 'Théoden Ednew', reinvigorated like Gandalf, capable as he dies of looking beyond death. Shire-poetry can be at once low mimetic and mythic, depending on whether one thinks of real forests, or forests as an image of life in the world. The flexibility with which Tolkien moves between the modes is a major cause for the success of *The Lord of the Rings*. It is at once ambitious (much more so than novels are allowed to be) and insidious (getting under the guard of the modern reader, trained to reject, or to ironize, the assumptions of tragedy or epic). This is how Tolkien in the end solved the problem first set up in *The Hobbit*, of bringing together the modern world of the Shire and the Bagginses, and the heroic world of were-bears and dragons and Thorin Oakenshield.

Literary mode is matched, of course, by style, and Tolkien's stylistic levels go up and down in exactly the same way as his generic ones. At the top we have the eagle's 'psalm' announcing the fall of Sauron; at the bottom, perhaps, the orcs, or Gollum/ Sméagol talking to himself. Most of Tolkien's variations from the middle level of the bourgeois novel have annoyed commentators, unused as they often are to earlier literature, or to contemporary popular literature. Tolkien has been criticized for writing a '*Boy's Own*' style, by Edwin Muir once again, and oddly enough, long ago, by Terry Pratchett, in the *Bath and West Evening Chronicle* for 7th December, 1974. It is true that hobbitic banter does sometimes sound like old-fashioned British school stories, now revived of course by the unexpected success of J.K. Rowling's 'Harry Potter' series. Tolkien has also been condemned for writing archaically (as he does, obviously deliberately, in scenes operating at the level of high mimesis or romance). There is a kind of presumption, however, in literary critics, usually utterly ignorant of the history of their own language, telling Tolkien what to think about English. Tolkien could at any time, and without trying, have rewritten any of his supposedly archaic passages either in really archaic language, in Middle English or Old English, or in completely normal demotic contemporary slang. In a letter composed for (but not sent to) a friend (Hugh Brogan) who made a complaint of this kind Tolkien in fact carried out just that exercise, rewriting Théoden's short speech beginning 'Nay, Gandalf!' in 'The King of the Golden Hall', first with a kind of advanced archaism including the old prefixed negative and second-person 'thou' [so 'You know not' becomes 'Thou n[e] wost'), and then in modern:

> 'Not at all my dear Gandalf. You don't know your own skill as a doctor. I shall go to the war in person, even if I have to be one of the first casualties ...'

'And then what?' Tolkien asked. How would a modern person, talking like that, express Théoden's heroic sentiment, 'Thus shall I sleep better'? As Tolkien replied:

people who think like that just do not talk a modern idiom. You can have 'I shall lie easier in my grave', or 'I should sleep sounder in my grave like that rather than if I stayed at home' – if you like. But there would be an insincerity of thought, a disunion of word and meaning. For a King who spoke in a modern style would not really think in such terms ... Like some non-Christian making a reference to some Christian belief which did not in fact move him at all.

(*Letters*, pp. 225–6)

Tolkien could bring a modern style into Middle-earth: Smaug talks it, for one, and so does Saruman. But he knew the implications of style, and of language, better and more professionally than almost anyone in the world. The flexibility of his many styles and languages; the resonance of the highest levels of these; the ability to reach out towards universal and mythic meaning, while remaining embedded in story: these are three powerful and largely unsuspected reasons for the continuing appeal of *The Lord of the Rings*.

CHAPTER V

❖─══◎═══─❖

THE SILMARILLION:
THE WORK OF HIS HEART

Lost lore and lays

The publication and success of *The Lord of the Rings* in 1954–5 left Tolkien in much the same position as the publication and success of *The Hobbit* in 1937. The publishers wanted a sequel, and this time they were seconded, as Stanley Unwin's son and successor Rayner confirms in a 1995 memoir, by increasingly large numbers of devoted readers. But Tolkien had no sequel ready to hand, or even in mind. What he had was what would be called nowadays (it is a word he would have hated) a 'prequel': the 'Silmarillion', existing as many manuscripts in many forms. He was never able to prepare this material for publication in a way which completely satisfied him, though he continued working on it for almost twenty years until he died; all the 'Silmarillions' now in existence have been published posthumously. Nevertheless it was the work of his heart, which occupied him for far longer than *The Hobbit* or *The Lord of the Rings*. The better-known works are in a way only offshoots, side-branches, of the immense chronicle/mythology/legendarium which is the 'Silmarillion', and which we have first in the form in which it was published as a connected narrative in 1977 (which I distinguish as *The Silmaril-*

lion), and then in many of the twelve volumes of *The History of Middle-earth* published between 1983 and 1996, all thirteen works (as also the volume of *Unfinished Tales of Númenor and Middle-earth*, from 1980) edited by Tolkien's son and literary executor Christopher.

Tolkien was working on something which might be seen as the seed of a section of *The Silmarillion* at least as early as 1913, when he began to write 'The Story of Kullervo', a 'prose-and-verse romance' never yet published which resembles in outline the story of Túrin, eventually chapter 21 of the 1977 *Silmarillion*. In late 1916, by now on convalescent leave from the trench fever contracted on the Somme, he was writing a much more extended and continuous account of elvish story, completed (or at least relinquished) by 1920, and published in 1983–4 as the two-volume *Book of Lost Tales*. During his years at Leeds University (1920–25) he began to versify two main sections of this material, the tales of Túrin and Beren, eventually published as *The Lays of Beleriand* in 1985. In 1926, when he sent one of these poems to his old teacher R.W. Reynolds, Tolkien also wrote a brief outline or 'Sketch of the Mythology' to act as background for Reynolds, which appeared as 'The Earliest Silmarillion' in *The Shaping of Middle-earth* (1986), though as with so much of what he wrote the published version takes in heavy rewriting up to 1930. Between 1930 and 1937, when *The Hobbit* came out, the 'Sketch' was rewritten in expanded form as the 'Quenta' or 'Quenta Noldorinwa', and then rewritten again as the 'Quenta Silmarillion' (the first published in *The Shaping of Middle-earth*, above, the second in *The Lost Road*, in 1987). It was this latter work, in its original form 'a beautiful and elegant manuscript', which was sent to Stanley Unwin in 1937 along with the poem 'The Gest of Beren and Lúthien' (the longest of *The Lays of Beleriand*), as a possible successor to *The Hobbit*. It met a confused reception from the publisher's reader, who seems to have seen only the poem and the section of the prose 'Quenta Silmarillion' added to explain it, and was gently rejected by Stanley Unwin as 'a mine to be explored in writing further books like *The Hobbit* rather than a book in itself' (see *The Lays of Beleriand*, pp. 364–7). Tolkien turned away to write *The Lord of the Rings*; but once this was over, first after the completion of *The Lord of the Rings* in 1951, and then after its publication in 1955, he turned back to rewrite his material once more, this time as 'The Later Quenta Silmarillion', published in two 'phases' in volumes X and XI of *The History of Middle-earth*, *Morgoth's Ring* (1993) and *The War of the Jewels* (1994). Even this account very much understates the complexity of *The Silmarillion*'s development (there is a much more extended account by Charles Noad, see 'List of References' below), for the versions listed above were often written and rewritten, in some cases becoming 'a chaotic palimpsest, with layer upon layer of correction'; while Tolkien also produced several sets of 'Annals' covering the same material, some of them written in Old English: 'The Annals of Valinor', 'The Annals of Beleriand', 'The Grey Annals', 'The Annals of Aman', all to be found in volumes IV-V and X-XI of *The History of Middle-earth*.

Generalizing successfully about this mass of heterogeneous material, a 'fixed tradition', as Christopher Tolkien notes, but never a 'fixed text', may seem to be impossible: but some shafts, one hopes of light, may be driven into it. One by this time predictable point is that Tolkien derived some part of his invention, and a vital part, from an entirely novel solution to an old mythological problem. There was no doubt that a belief in 'elves' (Old Norse *álfar*, Old English *ylfe*) was widespread in Germanic antiquity: but the words used about them seemed curiously contradictory. The Icelander Snorri Sturluson, whose prose account of Norse mythology remains our only half-coherent account, was aware of both 'Light-elves' (*ljosálfar*) and 'Dark-elves' (*dökkálfar*), but he also recognized 'Swart-elves' (*svartálfar*), though the place they lived, 'Swart-elf-home' (*Svartálfaheim*) was also the home of the dwarves. Meanwhile Old English uses words like 'Wood-elf' (*wuduælf*) and 'Water-elf' (*wæterælf*). How are all these fragments to be fitted together? Are 'Swart-elves' the same as 'Dark-elves', and both perhaps the same as dwarves? The *OED* seems to accept this solution, cross-referring 'dwarf' and 'elf' rather vaguely to each other, but it is a feeble notion. Early accounts distinguish the two species from each other perfectly clearly, the dwarves being associated with mining, smithcraft, and a world underground, the elves with beauty, allure, dancing, and the woodland. Tolkien's great predecessor, Jacob Grimm, also pondered the problem, but in his *Deutsche Mythologie* (translated into English in 1884 as *Teutonic Mythology*) could come to no conclusion, ending another rather vague discussion with the once-more feeble remark that maybe 'Dark-elves' were sort of in between 'Light-elves' and 'Swart-elves', 'not so much downright black, as dim, dingy'. It would be surprising if Tolkien had not read this passage in youth, and been annoyed by it.

At the heart of his account of the elves is a quite different distinction. The elves are not separated by colour (black, white, and 'dingy'), but by history. The 'Light-elves' are those who have 'seen the Light', the Light of the Two Trees which preceded the Sun and Moon, in Aman, or Valinor, the Undying Land in the West; the 'Dark-elves' are those who refused the journey and remained in Middle-earth, to which many of the Light-elves however returned, as exiles or as outcasts. The Dark-elves who remained in the woods of Beleriand are also, of course, naturally described as Wood-elves. And as for the connection with dwarves, the two species in Tolkien are quite distinct and never mingle, but they do in some cases associate. Elves may live underground and be given admiring dwarvish names, like Finrod 'Felagund' (< Dwarvish *felak-gundu*, 'cave-hewer'). It would only be natural, as time went by and memory became blurred, for men to be unsure whether such a character was once elf or dwarf, or what was the difference. A main aim in Tolkien's creations was always to 'save the evidence', to rescue his ancient sources from hasty modern accusations of vagueness or folly. Saving the evidence, moreover, generated story, in this case the complex story of the elves' wanderings, separations, and returns, summed up as well as anywhere in chapter 8 of *The Hobbit*:

> For most of [the Wood-elves] . . . were descended from the ancient tribes that never went to Faerie in the West. There the Light-elves and the Deep-elves and the Sea-elves went and lived for ages, and grew fairer and wiser and more learned, and invented their magic and their cunning craft in the making of beautiful and marvellous things, before some came back into the Wide World. In the Wide World the Wood-elves lingered in the twilight of our Sun and Moon, but loved best the stars; and they wandered in the great forests that grew tall in lands that are now lost.

Two less easily graspable points about *The Silmarillion* tradition concern language, and nationality. Tolkien said, in many ways, as forcefully as he could, and perhaps with a certain defensiveness (for writing fairy-stories was certainly seen by some in authority as a distraction from his proper job of being a language professor), that all his work was '*fundamentally linguistic* in inspiration' (his emphasis). The 'authorities of the university' might well consider his fiction a hobby, more or less pardonable, but to him it was not a hobby 'in the sense of something quite different from one's work, taken up as a relief-outlet'. Instead, 'The invention of languages is the foundation. The 'stories' were made rather to provide a world for the languages than the reverse' (*Letters*, p. 219). Accordingly, one might well say that the major root of *The Silmarillion* and all that followed from it was the invention of the elvish languages, Quenya ('Elf-latin', the language of the Light-elves), and Sindarin, the language of Beleriand, of the Wood-elves. One would get even closer by saying that the real root was the relationship between them, with all the changes of sound and semantics which created two mutually-incomprehensible languages from one original root, and the whole history of separation and different experience which those changes implied. (The best discussions of these are to be found on Carl Hostetter's web-site in the 'List of References' below). Such developments were Tolkien's major professional field, like those which generated (for instance) Gothic, Norse and English from one original root, perhaps preserved in the early runic monuments of Scandinavia. He himself suggested that the Quenya/Sindarin relationship was more like that between Latin and Welsh, though there is probably no one alive with the knowledge to appreciate it. Still, recondite though Tolkien's linguistic interests were, he could claim, and several times did claim in the lectures of his later years, that he had made his point by demonstration if not by argument: rooting story in language had *worked*, even for those who did not care, or did not know they cared about language.

Tolkien's views on nationality may be even more idiosyncratic, though they are straightforward and logical. His family name was German-derived, as he knew perfectly well, being a re-spelling of the nickname *tollkühn*, 'foolhardy'. He saw himself, however, as being 'far more of a Suffield (a family deriving from Evesham in Worcestershire)', and being like his family 'intensely English (not British)'. The trouble with this, as no one was in a better position to appreciate, was that native English tradition, following the Norman Conquest and the take-over by French and Latin learning, had in England been very largely, if not quite completely, suppressed. Jacob and Wilhelm Grimm, searching for relics of their country's past in the children's tales of the nineteenth century, had come up with quite a respectable haul, as had their followers recording Gaelic, or Irish, or Welsh. But the obverse of the domination of English as a language within the British Isles and elsewhere was that it had become international, multi-cultural, the language of the educated with no time for fooleries and fairy-tales. As a result, in spite of the very early start made on literacy by the Christianized English, native tradition petered out. The Welsh continued to tell tales of King Arthur, but there are (almost) no native stories of Hengest and Horsa; nineteenth-century English fairy-tale collections are among the weakest in Europe.

One thing Tolkien accordingly set himself to do – it will be remembered that his first extensive composition was called *The Book of Lost Tales* – was to reverse this decline, and restore to England something like the body of lost legend which it must once have had. His project is discussed by Carl Hostetter and Arden Smith in the volume of *Centenary Conference Proceedings*, but one may say briefly that this is why Tolkien spent much effort in writing 'The Annals of Beleriand' and some of 'The Annals of Valinor' in Old English: to provide a chain of communication between the imagined far past and the first beginnings of English history. It was, on a larger scale, the same sort of activity as that mentioned on p. 26 above, writing an Old English riddle to act as a reconstructed ancestor for modern nursery-rhyme. In early versions of the 'Silmarillion', furthermore, the stories are passed on by an early Englishman, or 'Anglo-Saxon', who has been stranded among the elves and learned their history directly from them. He, Eriol, or Ælfwine, is accordingly a witness to 'the true tradition of the fairies' (*The Book of Lost Tales II*, p. 290), not the 'garbled things' told by other nations. Tolkien's painstaking attempts to develop this idea have been discussed most recently by Verlyn Flieger, in her article 'The Footsteps of Ælfwine' in the recent collection *Tolkien's 'Legendarium'*. However, Tolkien also did rather more than flirt with the idea – though in the end he found it untenable – that Elvenhome had survived as England, with England as formerly Tol Eressëa, the Lonely Isle, Warwick as the elvish city Kortirion, and the Staffordshire village of Great Haywood, where he spent some of his convalescence, as Tavrobel, where Eriol learned the 'lost tales' of elvish myth (*The Book of Lost Tales I*, pp. 24–5). The theory could not work. For one thing, as Tolkien knew perfectly well, the English were themselves immigrants, who had come into Britain (not at that time Eng-land, the land of the English, in any sense) some fifteen hundred years ago; and though like the hobbits in the Shire they 'fell in love with their new land' and indeed forgot that they had ever had another one, it was impossible for a real historian to imagine a continuous tradition, in the same place, lasting from before the Romans and the ancestors of the Welsh through to the arrival of those whom Tolkien reckoned as his ancestors. Nevertheless, Tolkien would have liked to create a 'mythology' for his own people, to anchor it in the counties of the West Midlands, and simultaneously to preserve in it what scraps remained of the myths and legends there must once have been.

One of the clearest signs of Tolkien's overall intention is his use of the word 'lays', as in *The Lays of Beleriand*. 'Lay' is now an unfamiliar term with no precisely accepted meaning, just an old word, it seems, for 'poem'. Tolkien, however, did not think of it like that. What he did mean by it can be seen in another famous work from a century before his time, Lord Macaulay's *The Lays of Ancient Rome*. Many people have come across at least one of the poems from this set, the famous 'Horatius', which tells the story of 'Horatius at the Bridge', but few readers nowadays realize what Macaulay intended to do in it. Macaulay's 'Preface' makes it quite clear. Before his time (1842), one could say that works like Livy's *Roman Histories* had been perfectly familiar, indeed set reading for centuries of schoolboys. They had been accepted, though, just as histories, and even if one had one's doubts about the stories they contained, there seemed no prospect of correcting them or getting behind them to whatever earlier sources Livy must have used (all of them long vanished from the world). With the coming of the 'higher criticism' in Germany, however, methods were developed (largely subjective, but sometimes linguistic) for disentangling earlier from later strata of story, and genuine old tradition from contemporary faking. It became widely believed that behind the extensive epics of Homer, and Virgil, and the *Histories* of Livy, and *Beowulf,* and even the accounts of the Old Testament, there must have been early pre-literate traditions which were used by the later writers – traditions probably expressed in short poems composed at or near the time of the events they commemorated. Germans called these almost entirely hypothetical poems *Lieder,* while English-speaking authors divided between calling them 'ballads' and calling them 'lays': the *OED* defines 'lay', in this technical sense, as:

> the appropriate term for a popular historical ballad such as those on which the Homeric poems are supposed by some to have been founded. Some writers have misapplied it to long poems of epic character such as the Nibelungenlied or Beowulf.

It is clear that the *OED* editor here is unconvinced by the whole theory, with his 'misapplied' and 'supposed by some', but Lord Macaulay at least believed it. In the' Preface' to his *Lays of Ancient Rome* he put forward the argument that Rome, like England, had had a native tradition of balladic verse; but, like England, had been intellectually colonized by a culture felt by the educated classes to be superior (Greek culture for Rome, French culture for England); and had accordingly suppressed or abandoned its deepest roots. In England and Scotland (Macaulay said) the position had been partly rescued at almost the last possible moment by the activity of antiquarians like Thomas Percy and Sir Walter Scott, but Rome had not been so lucky. Still, something like the ballads of the Anglo-Scottish border must have existed in ancient Rome; that 'lost ballad-poetry of Rome' must have been transformed by the likes of Virgil and Livy into epic and history; and 'To reverse that process, to transform some portions of early Roman history back into the poetry of which they were made, is the object of this work.' So Macaulay wrote, not only 'Horatius' but three other ballads, 'The Battle of the Lake Regillus', 'Virginia', and 'The Prophecy of Capys'.

One attraction of this process, furthermore, and another one which could not have been aimed at without the work of the 'higher critics', was that even in these supposedly early ballads one could see some indication of date. German critics had become extraordinarily astute (usually too astute) in picking out anachronisms within the works they studied. They could tell (or thought they could tell) the difference between material which was original, which went right back to whatever historical event was being commemorated – in the case of *Beowulf*, say, the death of Beowulf's uncle in battle in the early sixth century – and material being inserted maybe two hundred years later, like the many Christian references in *Beowulf* which could only come from a time after the English had become Christians. One very bad result of this, to which Tolkien put a firm and complete stop with his 1936 lecture, was that much of *Beowulf* was effectively thrown away as 'phoney' by over-astute dissectors. But one good result was that people learned to read histories and historical poems with a kind of double vision, to see both the event being described and the context in which it was described. Macaulay built this kind of vision into 'Horatius' (and pointed out that he had done so in the 'Preface') by including evidently nostalgic remarks about 'the brave days of old', which show that his feigned 'lay' is deliberately looking backward from some historical distance. It has two dates in it, event, and record. This is the kind of thing flattened out by the treatments of Virgil or Livy.

'Alas for the lost lore, the annals and old poets that Virgil knew, and only used in the making of a new thing!' wrote Tolkien in his discussion of *Beowulf* (*Essays*, pp. 27–8), and in that context what he meant above all was that one should concentrate on the 'new thing' (the poem that survived) and not mope after all the hypothetical ones that hadn't. Just the same, he meant the 'alas' as well, felt the tragedy of the 'lost lore and annals', wished above all to create the sense of age, of antiquity with yet greater antiquity behind it, which was theorized by the 'higher critics', counterfeited by Lord Macaulay, and which Tolkien in his turn thought he could recognize in poems like *Sir Gawain and the Green Knight*. It was his quest for 'this flavour, this atmosphere, this virtue that such *rooted* works have' (*Essays*, p. 72), which led Tolkien to spend so much time and effort in creating different sets of 'annals', in different languages, imagined as the sources for the 'Silmarillion'; why the 1977 *Silmarillion* itself is studded with references to poems on which the imagined compiler is drawing, 'the *Noldolantë* ... that Maglor made before he was lost', the 'Lay of Leithian', the '*Laer Cú Beleg*, the Song of the Great Bow', to name only three of many; and why in the end and in some cases, like the 'Lay of Leithian', he went on to write the 'lay' itself, creating his own historical tradition like Lord Macaulay.

The effort of doing all this was extremely great, and the returns perhaps for many people rather minimal, for the sense of depth and age, the ability to read a work on two chronological levels at once, are rather recondite matters. Still, the sense that there was a deep and old tradition behind it, surfacing in the poems recited by Aragorn or Bilbo or Sam Gamgee, had been a major part of the texture even of *The Lord of the Rings*. Perhaps that success could be repeated in the 'Silmarillion'. In any case, one may feel, this was what Tolkien all his life most wanted to do. The first poem he ever published, in the *King Edward's School Chronicle* for 1911, was an account of an inter-house rugby match, called 'The Battle of the Eastern Field' (the school rugby pitches being off Eastern Road). As Jessica Yates has pointed out, it is quite clearly written, if in mock-heroic spirit, in a style which closely imitates Macaulay's *Lays*.

A parallel mythology

Commenting briefly on a tradition as complex and developed as this one is hard to manage with perfect accuracy. However, it can be said that in broad outline Tolkien's image of the history of the First Age remained relatively stable. It can be divided, if arbitrarily, into three main sections, indicated here according to the titles and chapter-numbers of the published 1977 *Silmarillion*.

The first 'section' consists of the 'Ainulindalë', the 'Valaquenta', and chapters 1–2 of the 1977 'Quenta Silmarillion' itself. They deal with the creation of the world, the rebellion of one of the Creator's subordinate spirits, Melkor, and the decision by some of these subordinate spirits (the Valar), including Melkor, to bind themselves within the world, Earth (though to them Earth includes Aman, the Undying Lands, as well as Middle-earth, the lands of mortality).

Chapters 3–8, and 11, deal with the appearance of the elves, the decision of the other Valar to imprison Melkor in order to protect the elves, the migration of the Light-elves from Middle-earth to Aman, and the unrest and destruction caused there by the released Melkor and an elvish faction. Here the 'Silmarils' appear. They are jewels made by the greatest of the elvish smiths, Fëanor, and they contain within them the light of the Two Trees of Valinor, the trees which lit the world before the rising of Sun and Moon. Once the Trees are poisoned by Melkor and his spider-ally Ungoliant, their light survives only in the Silmarils. But when Fëanor is called on to give them up, to be broken to bring the Trees back to life, he refuses – only to find that Melkor has already stolen them. Fëanor, his sons and his adherents (mostly from his own tribe of the Noldor) then decide to leave Aman to pursue Melkor and regain the Silmarils, and he and his sons swear an oath of vengeance on anyone, 'Vala, Demon, Elf or Man as yet unborn ... whoso would hold or take or keep a

Silmaril from their possession'. In pursuit of this oath they commit two initial acts of violence or treachery: they steal the ships of the elves who dwell on the shores of Aman (the Teleri), killing many in the process; and then having crossed back to Middle-earth, from which they had been brought by the Valar, they burn the ships and refuse to return for those of their supporters (including Galadriel) left behind. The latter group reach Middle-earth only by marching across the ice of the North. Meanwhile the Valar, dismayed by the loss of the Trees and the defection of Fëanor, create the Sun and Moon to replace the Trees, but cut off communication between Middle-earth and Aman.

The third and longest section, effectively chapters 10 and 12–24 of the 1977 *Silmarillion*, deals with the wars in Middle-earth between the elves and Melkor (renamed Morgoth), and the ill-fated attempts to regain the Silmarils. Into these wars are drawn both men, who appear in chapter 12, and dwarves, while Morgoth deploys orcs, balrogs, and dragons. They are marked also by internal feuding and treachery, while the two longest chapters deal with the human heroes Beren, who regains a Silmaril at the cost of his hand, and Túrin the ill-fated: it is these stories which Tolkien versified in *The Lays of Beleriand*. The Silmaril recaptured by Beren goes from one owner to another, always bringing disaster with it. In the end Eärendil, a hero of mixed elvish/human ancestry, sails with its aid to Aman, to beg the Valar in Valinor for forgiveness and assistance to Middle-earth. This is granted; Morgoth is overthrown; and the two remaining Silmarils are recaptured, only to be lost again by the final workings-out of the oath of Fëanor and his sons. Eärendil's Silmaril, however, shines from the prow of his ship, which has been set in the sky as a star and a sign of hope to Middle-earth.

Even from this summary one can see several things. *The Silmarillion* bears a kind of relationship to Christian myth. The rebellion of Melkor, and his subordinate spirits, is analogous to the Fall of Lucifer and the rebel angels. Lucifer is by tradition *princeps huius mundi*, 'the prince of this world', and Melkor calls himself, perhaps truthfully, 'Master of the fates of Arda'. The origin of the fall is also the same in both cases, for the sin of Lucifer was (according to C.S. Lewis) the urge to put his own purposes before those of God, and that of Melkor was 'to interweave matters of his own imagining' with the 'theme of Ilúvatar [the Creator]'. This 'fall of the angels' also leads in both mythologies to a second fall: the Fall of Man and the exile from the Garden of Eden in the *Book of Genesis*, the loss of elvish innocence and the emigration from Aman (which becomes an exile) in *The Silmarillion*. Finally Eärendil, the half-human emissary who obtains forgiveness and rescue from the Valar, bears a more distant similarity to the Incarnation of Christ, and the Christian promise of salvation.

That said, there are of course very marked differences, especially centring on the Silmarils, to which there is no close Christian parallel. These differences shed more light on the nature of *The Silmarillion*, for after all it would seem pointless – it might even be thought presumptuous – just to rerun the Christian myth, in which Tolkien devoutly believed, merely in a work of human imagining. There is in fact an ambiguity running through *The Silmarillion*, which is this. The four most powerful of the Valar are clearly the spirits of earth, water, air and fire, respectively Aulë, Ulmo, Manwë, and Melkor: Melkor is the spirit of fire. Meanwhile Fëanor, the maker of the Silmarils and the elf responsible for the second fall, is actually a nickname and means again 'Spirit of Fire'. But Fëanor is an ambiguous character, proud, selfish, vengeful – but also skilled, ambitious, demanding justice. At the heart of his fall is the refusal to give up the works of his own craft, crying out bitterly, when he is asked to surrender the Silmarils:

> 'For the less even as for the greater there is some deed that he may accomplish but once only; and in that deed his heart shall rest.'
>
> (*The Silmarillion*, chapter 9)

Tolkien clearly had more than a certain sympathy with this view. It was something he felt himself (only his Silmaril was *The Silmarillion*). In his essay 'On Fairy-Stories' he protested strongly, even passionately, that there was a right to create fantasy, even if, even though, fantasy could be abused, could become the making and worshipping of false gods, whether literally (like Beelzebub, the 'Lord of the Flies' of Golding's fantasy), or politically, in the shape of 'social and economic theories' also demanding 'human sacrifice'. But fantasy was a human desire which could not be taken away:

> At the heart of many man-made stories of the elves [surely, in this case, *The Silmarillion*] lies, open or concealed, pure or alloyed, the desire for a living, realized sub-creative art, which (however much it may outwardly resemble it) is inwardly wholly different from the greed for self-centred power which is the mark of the mere Magician [or, one might say, the 'Necromancer', Sauron]. Of this desire the elves, in their better (but still perilous) part are largely made . . .
>
> (*Essays*, p. 143)

The 'sub-creative' desire, then, is legitimate – Tolkien goes on to say, in a fragment of verse, it is 'our right . . . That right has not decayed: / we make still by the law in which we're made'.

But if it is legitimate for Tolkien, is it for Fëanor? And how does it relate to the urge to manufacture which creates not only the Silmarils but also the invention of weapons:

> Fëanor made a secret forge, of which not even Melkor was aware; and there he tempered fell swords for himself and for his sons, and made tall helms with plumes of red.
>
> (*The Silmarillion*, chapter 7)

240

One Old English poem, which Tolkien certainly knew (it gave him the word *éored*), seems to locate the Fall of Man not in Adam and Eve and the Garden of Eden, but in Cain and Abel and the invention of metallurgy: 'A state of violence came into being for the race of men, from the moment when the earth swallowed the blood of Abel ... So the inhabitants of Earth endured the clash of weapons widely through the world, inventing and tempering wounding swords'. And one might recall that Fëanor is not the only dangerous maker in Tolkien's work. Saruman too is a forger and creator and a user of fire, whose name could indeed be translated 'Artificer', or even 'Engineer', see pp. 169–71 above. Thorin Oakenshield's disastrous fascination with the Arkenstone parallels the disastrous quests for the Silmarils, but is also only a normal dwarvish urge raised to a higher power, the urge even Bilbo feels for a moment, 'the love of beautiful things made by hands and by cunning and by magic ... the desire of the hearts of dwarves'. In *The Silmarillion* too Aulë, the earth-spirit of the Valar and patron of 'all craftsmen', makes the dwarves against the wishes of Ilúvatar, and weeps when he is detected and must offer to destroy them. Tolkien's work in fact gives a continuum of creative urges, from wholly evil (Melkor's, the 'self-centred' urge of the 'Magician'), to wholly legitimate (his own, the right to fantasy and to 'sub-creation'); but they shade into each other, and it is not easy to see always why they should be distinguished. Some of the tension of *The Silmarillion* comes from sympathy with the sons of Fëanor and their disastrous oath, and with those who reject Aman and unchanging immortality for Middle-earth, creation, independence, and death. And, one might add, entirely seriously, for linguistic change, which happens only in Beleriand: Tolkien once applied the term *felix peccatum*, 'fortunate sin', not to the Fall of Man (which was made 'fortunate' by the Incarnation) but to the Tower of Babel, the presumption which by tradition created the multiplicity of human languages from a single root (see *Essays*, p. 194).

Tolkien indeed built the concept of the *felix peccatum* into his own mythology, when Ilúvatar declares that even Melkor in his sin 'shall prove but mine instrument in the devising of things more wonderful'. And in a sense he also built in, or rather left a space for, the traditional story of the Fall of Man. There is no Garden of Eden for humans in *The Silmarillion*, but when humans do enter Middle-earth from the east all that is known about them to the elves who are imagined as the preservers of these traditions is that something dreadful had happened to them already, a 'darkness' which 'lay upon the hearts of Men' and which was connected with an unknown expedition of Morgoth: one could believe that Morgoth here is identical with Satan, and his expedition was to lure humanity into their 'original sin'. *The Silmarillion* then does not contradict *Genesis*; but it does offer an alternative view of the origin of sin, in a desire not for the 'knowledge of good and evil', but in the desires for creation, mastery, power.

'A passion for family history'

These desires are then worked out in the long history of elves and humans, which occupies sections two and three of *The Silmarillion*, as divided above. Most people have found these hard to follow – the most cutting remark made being the apocryphal, 'a telephone directory in Elvish, yet' (for which I know no source). There are an awful lot of names in *The Silmarillion*, it is true, and the genealogies trespass on the short memories of modern literates. Nevertheless, a clear structure for the whole work can be made out if one once masters the central idea of the divisions among the elvish tribes.

The most basic of these is the distinction between the Light-elves, the Calaquendi, who reached Valinor and saw the light of the Two Trees before they were poisoned, and the Dark-elves, the Moriquendi, who refused the journey. (The divisions given here are necessarily something of an approximation. For a full picture, see the information laid out in diagrammatic form in the appendices to *The Silmarillion*.)

This latter group is largely the same as the speakers of Sindarin, the elvish language which developed by language-change in Beleriand, but not always and not absolutely. One of the three original ambassadors to Valinor was Elwë Singollo (in Quenya, Elu Thingol in Sindarin), who returned to Middle-earth to urge his people to go to Aman, but then remained behind himself, held by his love for Melian the Maia (Maiar are spirits intermediate between the elves and the Valar, and include both Gandalf and the Balrogs). He, accordingly, though king of the Sindarin-speaking Grey-elves, Elves of the Twilight, is not to be counted as a Dark-elf; for he had seen the light once. And yet he is called a Dark-elf, on one occasion. When the Noldor arrive in Middle-earth, Elwë is naturally wary of what may turn into a dispossession, and sends a message warning the sons of Fëanor to stay within the limits he has set. He has been informed of the situation by 'Angrod son of Finarfin', who is both Fëanor's nephew (on his father's side) and Elwë's great-nephew (on his mother's side). The sons of Fëanor take offence at this, and one of them, Caranthir, calls out:

'Let not the sons of Finarfin run hither and thither with their tales to *this Dark Elf in his caves*! Who made them our spokesmen to deal with him? And though they be come indeed to Beleriand, let them not so swiftly forget that their father is a lord of the Noldor, *though their mother be of other kin.*'

(my emphases, *The Silmarillion*, chapter 13)

This is an offensive speech on several levels, and the offence is compounded by containing half-truths. Elwë is not technically speaking a 'Dark Elf', for he has been to Valinor and seen the Light of the Trees; on the other hand he is king of the Dark-elves, and he refused to return to Valinor, so there is a basis for Caranthir's claim of superiority. As for the sons of Finarfin, Caranthir's sneer is based on the fact that their mother, Elwë's niece, though a Light-elf, comes from the most junior branch of the Light-elves, the Teleri. But the sneer could easily be turned back on him, for they are also descended, on their grandmother's side, from the Vanyar, the most senior branch (the Noldor, to which the sons of Fëanor belong, being intermediate). So one of Caranthir's claims is false in detail but true in general, while the other is true in detail and false in general. This is, then, a subtle and a tense situation, one of many which build up the overall effect of the tragedy of the Noldor. But the subtlety and the tension depend on carrying in one's head a string of distinctions between elvish groups, and a whole series of pedigrees and family relationships. The audiences of Icelandic sagas could do this, but readers of modern novels are not used to it, and easily miss most of what is intended.

Though Tolkien meant *The Silmarillion* to fill a gap in English tradition, he does indeed seem to have drawn mostly on Old Norse or Icelandic literature for its main themes. The Silmarils themselves, in my opinion, are an attempt to solve the mysterious riddle of the *sampo*, an undefined object often referred to in the Finnish *Kalevala* – Tolkien was fond of Finnish, modelled aspects of Quenya on it, and furthermore admired the *Kalevala* as a product of exactly the kind of literary rescue-project which he would have liked to see in England, see pp. xv, xxxiv above. However, much of *The Silmarillion* can be seen as a complex tragedy of mixed blood, of the kind seen in several poems of the *Elder Edda*. To revert to Dark-elves, the Dark Elf *par excellence* in *The Silmarillion* is Eöl, a great smith, a close associate of the dwarves (so answering one of the points bodged by Jacob Grimm, see p. 229), a relative of Elu Thingol and hostile to the incoming

Noldor. He captures and marries Aredhel, lost in the woods of Beleriand, and so sets up a sequence of tragedies – all dependent on genealogy. Aredhel is first cousin both to the sons of Finarfin and to the sons of Fëanor. She is also the sister of Turgon who, distrusting the confidence of the sons of Fëanor, has withdrawn into one of the three 'Hidden Kingdoms', Gondolin. Why does Aredhel leave Gondolin, and so start the tragic sequence? Her interview with her brother suggests both pride and deliberate deception – one of several snowball-that-starts-an-avalanche scenes in *The Silmarillion*. Be that as it may, she uses her permission to visit her Telerin/Vanyarin cousins, the sons of Finarfin, to try to reach her Noldorin cousins, the sons of Fëanor: an ominous choice. In any case she fails, is captured by Eöl, and bears him a son, Maeglin. They eventually escape into the country of the sons of Fëanor, who capture Eöl as he follows them. What is his relationship to them? He claims kinship, punningly and sarcastically, as Curufin expels him, 'It is good to find a kinsman thus kindly at need' – and this is as usual part-true, for he is a cousin by marriage. But Curufin rejects the claim and denies the connection: 'those who steal the daughters of the Noldor . . . do not gain kinship with their kin'. This scene is however contrasted with the next one, in which Eöl, still pursuing his wife and son, finds his way to Gondolin and is captured by Aredhel's brother Turgon. Turgon, by contrast with Curufin, magnanimously grants the connection, immediately greeting him with 'Welcome, kinsman, for so I hold you'. But Eöl (who still bears the grudge of the dispossessed Dark-elves) rejects the offer, demands his wife and son, and when this is refused kills Aredhel and is himself executed.

Both Eöl and Aredhel have in different ways been studies in ambiguity, but that ambiguity now shifts to their son Maeglin. His closest relative is now his uncle Turgon; his uncle however killed his father; his father however killed his mother; and behind all the 'howevers' there is the question of inheritance. Should

Maeglin inherit from his uncle? He is less than half Noldorin by blood. In any case, whatever his feelings about his father, he may have inherited from him the Telerin grudge, of dispossession. A solution would be for Maeglin to marry his cousin, Turgon's daughter and only child, Idril; but first-cousin marriage is forbidden in elvish society (though not in Tolkien's own society). When Maeglin sees himself displaced once more by the human Tuor, who marries Idril and fathers Eärendil, he turns traitor and betrays Gondolin to Morgoth. Who, then, is responsible for the Fall of Gondolin? Maeglin, for treason? Eöl, for abducting Aredhel? Aredhel, for defying her brother? The sons of Fëanor, for their pride, lack of respect for others, and bad example? The strains of mixed blood? Elvish historians, according to Tolkien, see the core of it in Maeglin's urge towards a kind of incest, and regard this as 'an evil fruit of the Kinslaying', the first Noldorin assault upon the Teleri in Aman: it is a sort of sexual retaliation for ancient violence.

The whole train of story, then, is a sad and complex one, with many mixed motives and scenes of hidden tension. But in order to follow it, it is vital to remember who everyone is, who their relatives are, and what they feel about their relatives. As has been said above, readers of Norse sagas could do it, and according to Tolkien, hobbits could do it too, as one sees from remarks like Gaffer Gamgee's 'So Mr Frodo is [Bilbo's] first and second cousin, once removed either way, as the saying is, if you follow me'. Few do follow him, though; and though the hobbits 'have a passion for family history', it is not always shared. The organization of *The Silmarillion* in that respect makes demands upon its readers which no other modern work has ventured, including (for all its complex structure) *The Lord of the Rings*.

The 'Human-stories' of the elves

Another way of penetrating the structure of the third section of *The Silmarillion* is to observe that it is largely organized round the falls of three different 'Hidden Kingdoms': Doriath, Nargothrond, Gondolin. Each is set up by an elf-king, respectively Elu Thingol, Finrod Felagund, Turgon, the latter two motivated by lack of faith in the power of the sons of Fëanor to ward off Morgoth (Thingol made his decision before the return of the Noldor to Middle-earth). Each kingdom prospers for a while, even a long while, till each is located by and willingly or unwillingly acts as host to a mortal, a man, respectively Beren, Túrin, Tuor. These stories of human involvement with the elves were above all the works of Tolkien's heart. Christopher Tolkien has noted his father's statement that the tale of Tuor and the fall of Gondolin was the first of the *Silmarillion* complex to be written, while on sick leave from the army in late 1916 or 1917 (*The Book of Lost Tales I*, p. 10). It leads furthermore to the story of Eärendil, Tuor's son, a name which we know had caught Tolkien's attention even earlier, while he was still a student at Oxford, and which had generated what is possibly the very first work in his whole mythological cycle, the poem of 'The Voyage of Earendel', written in September 1914 (*The Book of Lost Tales II*, pp. 267–9). Meanwhile the story of Beren and Lúthien remained deeply personal to Tolkien till he died: he had the names 'Beren' and 'Lúthien' carved on his and his wife's shared tombstone, a striking identification. These tales do indeed give a clue to the original motive and deepest theme of *The Silmarillion*, and perhaps of all Tolkien's work.

In his essay 'On Fairy-Stories' (first published in 1947), Tolkien remarked that the 'oldest and deepest desire' satisfied by fairy-stories is to tell tales of 'the Great Escape: the Escape from Death'. He added, with clear self-reference which must in 1947 have seemed merely jocose, 'The human stories of the elves are doubt-less full of the escape from deathlessness'. The only such stories, of course, are those written by Tolkien, and not surprisingly, they do contain both themes. Beren escapes from death – he dies, but is brought back from the dead, alone among men, by the songs of Lúthien which move even Mandos, keeper of the Halls of the Dead, to pity. Lúthien correspondingly escapes from death-lessness, for she, like Arwen in Appendix A of *The Lord of the Rings*, is allowed to choose death and finally accompany her husband. Eärendil and his wife Elwing in their way also escape from mortality, and reach the Undying Lands to beg for aid to Middle-earth. But again conversely, and on a much larger scale, one should note that nearly all Tolkien's elvish characters choose death in the long term (though to them death is differ-ent from what it is for humans), simply by returning to Middle-earth. Their return does not make them immediately mortal, but it does expose them to the malice of Morgoth and the chances of Middle-earth, which are almost invariably fatal. Why do they do it? Why did Tolkien even imagine such a strange motivation?

There is no difficulty in seeing why Tolkien, from 1916 on, was preoccupied with the theme of death, and escape from it. By the end of World War I, as he said himself, his closest friends were dead. He had been an orphan since his mother died when he was twelve, and had never really known his father, who died when he was four. The theme of escape from death might then naturally seem attractive. More puzzling is the theme of the escape to death, the deep love of the elves for the mortal world, which on the one hand they regard as *galadhremmin ennorath*, 'tree-tangled Middle-earth', and which they regard on the other as a paradise, loss of which is not even fully compensated by immortality, see Haldir's remark quoted on pp. 205–6 above. One might argue that Tolkien, elaborating his stories of a race choosing the fate of mortality, was trying to persuade himself that mortality had after all some attractions, invisible though those might be to humans who have no other choice. Against that, the whole of *The Silmarillion*, and especially the 'human-stories' embedded in it, is deeply sad, sad beyond *The Lord of the Rings* (though that is not as pain-free as imperceptive critics have said), certainly sad beyond anything normally tolerated in twentieth-century fiction. The question they ask insistently is 'why? Why do death and pain and evil come? Why are they necessary?'

Tolkien's answers to these perhaps unanswerable questions, long-evolving and never in fact completed though they are, can be seen in their most developed form in the tale of Túrin. This survives, like so much of *The Silmarillion*, in several major (and more minor) forms, which I would pick out as follows:

(1) the tale of 'Turambar and the Foalóke', in *The Book of Lost Tales II* (written by mid-1919)
(2) 'The Lay of the Children of Húrin', incomplete, written in alliterative verse in two main versions, in *The Lays of Beleriand* (written between 1922 and 1925)
(3) 'Of Túrin Turambar', chapter 21 of the 1977 *Silmarillion* (constructed from several sources, but perhaps predominantly work prior to 1937)
(4) the 'Narn î Hin Húrin', or 'Tale of the Children of Húrin', in *Unfinished Tales* (much the most expanded version as far as it goes, but fragmentary, written from 1951, see *The War of the Jewels*).

All four versions differ from each other, but the outline remains surprisingly stable.

Very briefly, the start of the story is the self-sacrificing stand of Túrin's father Húrin at the Battle of Nirnaeth Arnoediad, which allows Turgon to escape to Gondolin and puts Turgon under deep obligation. Húrin is taken alive by Morgoth and allowed to see the fate that unfolds for his children. Túrin's mother Morwen sends him away for safety, and he is received by Elu Thingol in Doriath. But Túrin, angered by taunts at his mother, kills one of the king's counsellors (Saeros), flees, and becomes an outlaw. Assisted by the marchwarden of Doriath, Beleg, who has remained his friend, he rises to prominence again, but is captured by the orcs, and on being rescued by Beleg, kills him by mistake. He makes his way to Nargothrond (where Finrod is now dead), and under a false name once more becomes prominent; he persuades the elves of Nargothrond to emerge from hiding and take a more aggressive role, while the new king's daughter, Finduilas, falls in love with him. Meanwhile Morwen, with her daughter Nienor, have eventually fled, and found refuge in Doriath, too late to catch up with Túrin. Túrin's new aggressive policy however only betrays Nargothrond to Morgoth, and it is destroyed by Glaurung the dragon. Glaurung's 'binding spell' holds Túrin immobile while Finduilas is driven away, and the dragon taunts him with abandoning his mother and sister. Túrin tries to rescue his mother, but arrives at their old home to find she has gone; and while he is doing that, Finduilas is killed by the orcs. Meanwhile Morwen, now looking in her turn for her son, meets the dragon, and is lost in the confusion, while Nienor, Túrin's sister, loses her mind, and runs naked through the forest till she collapses on the grave-mound of Finduilas. Túrin finds her there, and since neither he nor she knows who she is, marries her under the name of Níniel. In a last exploit, he wounds Glaurung mortally, but falls unconscious himself; and when Níniel comes to rescue him, Glaurung restores her memory to her, and she realizes she is pregnant by her brother. She commits suicide; Túrin kills the man who tells him what has happened, Brandir; but when it is confirmed, decides on suicide in his turn. In a last scene he asks his sword (the work of Eöl the Dark-elf) whether it will kill him, and it replies (in a scene which changes little from 1919 to 1951, and which is certainly imitated from the Finnish *Kalevala*):

'Yea, I will drink thy blood gladly, that so I may forget the
blood of Beleg my master, and the blood of Brandir slain
unjustly. I will slay thee swiftly.'

(*The Silmarillion*, chapter 21)

He kills himself, and the sword breaks as he does it.

All this is seen, furthermore, by Húrin, given the gift of vision
by Morgoth, and his embitterment once released plays a part in
the later destructions of both Doriath and Gondolin. But what
is the root of the tragedy? One answer is obviously that Túrin
brings his troubles on himself: again and again he lashes out and
kills the wrong person, Saeros, Beleg, Brandir, and others. Another could be that it is just terrible bad luck, if you believe
that luck is ever 'just' luck: Morwen and Túrin criss-cross while
looking for each other, and Nienor just happens to be found on
the grave of Finduilas, where Túrin's guilt and protective urges
are at their strongest. Or, of course, it could all be the fault of
Morgoth and his servant the dragon Glaurung, who spares Túrin
at Nargothrond only for a worse fate. But all three of these could
be seen as relatively comfortable explanations. As with the Eöl/
Aredhel/Maeglin complex discussed above, Tolkien remained
keenly interested in the hidden roots of evil or of disaster, in the
way that minor outbreaks of selfishness or carelessness mean
more than they seem: snowballs leading to avalanches once more.

These concepts are most developed in the latest version, the
'Narn i Hîn Húrin', incomplete though it is. The initial scene is
found in all versions. In *The Book of Lost Tales II* Melko (i.e.
Morgoth) curses Úrin (i.e. Húrin), putting 'a doom of woe and
a death of sorrow' on his family, and granting him 'a measure
of vision', so he can see what happens to them. The scene is there
in 'The Lay of the Children of Húrin' (*The Lays of Beleriand*),
where the phrase is 'a doom of dread, of death and horror'. In
the 1977 *Silmarillion* it has become 'a doom . . . of darkness and
sorrow', and here Melkor/Morgoth calls himself 'Master of the

fates of Arda' (i.e. Middle-earth). In the much-expanded 'Narn' version of this scene, Morgoth says further:

> 'I am the Elder King: first and mightiest of all the Valar, who was before the world, and made it. The shadow of my purpose lies upon Arda, and all that is in it bends slowly and surely to my will. But upon all whom you love my thought shall weigh as a cloud of Doom, and it shall bring them down into darkness and despair.'
>
> (*Unfinished Tales*, p. 67)

Húrin denies this, or some of it: 'Before Arda you were, but others also; and you did not make it'. And even if he was the mightiest of the Valar, Húrin adds, he could not pursue even mortals 'Beyond the Circles of the World'. Morgoth replies that there is nothing beyond the circles of the world. Húrin's last words are 'You lie', and Morgoth replies 'You shall see and you shall confess that I do not lie'. The question is, how far is Morgoth lying; and the fear is that some of what he says is true. Perhaps Morgoth really is (and Húrin makes allusive reference to whatever it was that caused the Fall of Man long ago, when Morgoth may have taken the role of Satan) the *princeps huius mundi*. Tolkien was after all in his own life intimately acquainted with 'the problem of pain', as Lewis called it. However, if the world is delivered over to a diabolic power, that power, it seems, must work through human wills, as the 'Narn' allusively suggests.

Some responsibility, to begin with, is laid in the 'Narn' on Túrin's mother Morwen. She is given very clear advice by her husband before he leaves, '*Do not be afraid!*' and '*Do not wait!*' She remembers this, but she ignores it, because 'she would not yet humble her pride to be an alms-guest', even of Thingol. She sends her son away instead, but it is a son who remembers the unfortunate words of his father's crippled servant Sador, that the incomers who take over the country have learned from the orcs to hunt their slaves with hounds. The fear that this may happen to his mother is clearly Túrin's major trauma – the image of a naked woman running. At Thingol's court it is the again unfortunate allusion to this – 'Do [the women of Hithlum] run like deer clad only in their hair?' – which triggers Túrin's first outbreak, first manslaughter, and second exile. The taunt which Glaurung levels at him (in the 1977 *Silmarillion*) is that he has abandoned his mother and sister, 'Thou art arrayed as a prince, but they go in rags', and it is his reaction to that which makes him abandon Finduilas to the fate which he fears for other women. But the fate he fears is exactly what comes about, with Morwen lost in the woods, and Nienor hunted by the orcs till she runs naked, 'as a beast that is hunted to heart-bursting'. It is pity for this, and the identification of all the abused women of his imagination in the one figure, which makes Túrin love Nienor, attempt to protect her by marrying her, and set up the fatal and final incest. All this comes from Morwen's bad decision to separate from her son, and one of its roots is pride.

It comes also from a series of (as I have called them above) 'unfortunate' phrases and allusions – none of them (except perhaps Glaurung's) intentional. But what is meant by 'fortune'? Is it the same as 'fate'? Túrin asks Sador 'What is Fate?' as a child, and gets no clear answer. But the implication of the story is that Morgoth was not lying, though he may not have been telling the whole truth, when he called himself 'Master of the fates of Arda'. He cannot make people do wrong, for that would deny human free will. But he can put words into their mouths, and the responses to those words, in the end all Túrin's fatefully bad decisions, are then their responsibility. Characters in Icelandic sagas occasionally say, of loose or provocative speech, 'trolls must have plucked at your tongues', and Tolkien repeats the idea in more dignified form, with Mablung saying for instance to Saeros after his taunting, 'some shadow of the North [i.e. of Morgoth] has reached out to touch us tonight'. Morgoth's 'doom', then, works by 'shadow' and suggestion. But *The Lord of the Rings* shows how shadow, absence, can paradoxically become a presence. In the 'Narn' the double explanations seen in *The Lord of the Rings* (see pp. 145–6 above) are strongly marked and openly discussed. Sador is lamed 'by ill-luck or the mishandling of his axe' (which? Morgoth could have sent the 'ill-luck', so that Sador would be there to say his 'unfortunate' words); Túrin reaches Doriath 'by fate and courage' (but the 'fate' may be Morgoth's, for it would be better if he had died young); Sador tells Túrin, like Galadriel talking to Sam Gamgee, 'a man that flies from his fear may find that he has only taken a short cut to meet it'. Túrin takes the nickname Turambar, 'master of doom', in defiance of this whole train of thought, asserting his own free will; but the epitaph given him by Nienor is *Túrin Turambar turún' ambartanen*, 'master of doom by doom mastered'. The whole story suggests once more a deep consideration of the nature of *Macbeth*, where similarly the words of the witches seem to bring about Macbeth's fall, but could not operate without Macbeth's responses to them.

A case could be made for seeing Tolkien's other major 'human story', the tale of Beren and Lúthien, as the philosophical antithesis to Túrin. It is a story of love across the species of elf and human, rather than a tale of incest; it contains the defeat of Morgoth and the recovery of a Silmaril, not the fulfilling of his purpose; it leads on to a more sustained triumph yet, for the couple's granddaughter is Elwing, the wife of Eärendil, who brings the Valar back to Middle-earth, while Túrin has no descendants; Lúthien masters fate and death in a way that Túrin cannot even aspire to; and the last word of the poetic version of the story sung by Aragorn in *The Fellowship of the Ring* (I/11) is 'sorrowless'. However, though the story contains both the Escape from Death (for Beren, sung back from the dead by his wife), and the Escape from Deathlessness (for Lúthien, given permission to join her husband in the end and pass beyond 'the Circles of the World'), it does not read like a 'comedy' in even the Dantesque sense. Work of Tolkien's heart though it was, and existing in even more versions than Túrin – including Aragorn's poem and an early version of that poem published separately in 1925 – the impression that it makes in the end remains one of crowding, and perhaps of derivation. It is full of motifs taken from earlier story – werewolves and vampires, the healing herb, the rope of hair let down from a window (as in 'Rapunzel'), the wizards' singing-contests (from the Finnish *Kalevala*). Its core may be Beren's rash promise to Thingol, 'when we meet again my hand shall hold a Silmaril', to be proved true in the letter, if false in the spirit, when Beren shows Thingol the stump of his arm, bitten off at the wrist by Carcharoth the wolf, so that his hand with the Silmaril in it is still in the wolf's belly. Yet Rash Promises between mortals and the inhabitants of Faerie are an old tradition, as in *Sir Orfeo* (where a mortal rescues his wife from the underworld through the fairy-king's promise), and *Sir Gawain and the Green Knight* (where the mortal makes the promise to receive a return-blow); and the motif of the wolf-bitten wrist is one of the most familiar tales from Snorri Sturluson's *Prose Edda*, where it is told of the god Tyr and the monster-wolf Fenrir.

Tolkien was happy, of course, to use old motifs and make them familiar once more; and my earlier criticism (in *The Road to Middle-earth*) of crowding, of the tale in the 1977 *Silmarillion* version feeling like a compendium, has been answered by Christopher Tolkien, who points out that that is exactly what the 'Silmarillion' at an earlier stage was intended to be: a compendium of legend made at the end of the Third Age of Middle-earth, abbreviating much earlier material but using some passages from it, like the quotation from the 'Lay of Leithian' given in *The Silmarillion* chapter 19 (*The Book of Lost Tales II*, p. 57). To return to the comments on Lord Macaulay made earlier, Tolkien's 'Quenta Silmarillion' of 1937 could readily be seen as taking the role of Livy's *Histories*, so to speak, while *The Lays of Beleriand* behind it, including 'The Lay of Leithian', would represent the 'lost lore' and the 'old poets' later used 'in the making of a new thing'. Tolkien's literary intention is then perfectly clear, and admirably consistent over a long period of time, as are his thematic concentrations on death and immortality, sorrow and consolation. One can see, though, that none of this had any connection at all with any literary mode now familiar, still less commercially viable. Stanley Unwin's gentle hint that the 'Silmarillion' material should be used in the writing of new *Hobbits* was no doubt well meant, but one cannot imagine how it could ever have been taken. Bilbo can co-exist with Thorin Oakenshield, and Frodo with Strider, but reducing Túrin and Beren to the 'low mimetic' mode of the modern novel would present a seemingly unscalable challenge.

Angels and the evangelium

At the end of all versions of *The Silmarillion* comes the tale of Eärendil, another story of escape from Middle-earth, a story which blends the escapes of human and elf, Eärendil being descended from both. In it Tolkien leaves the mode of heroic chronicle and returns to that of mythology. Like Saint Brendan, Eärendil continually explores to the west in search of the Undying Lands. In his absence the remaining sons of Fëanor attack his settlement in the hope of winning back the Silmaril inherited by his wife Elwing from her grandmother Lúthien. They fail, and Elwing throws herself and the jewel into the sea. But (and here the mythological strain becomes dominant once more) the Valar turn her into a bird, in which shape, and still carrying the Silmaril, she rejoins her husband far out at sea. The Silmaril takes them through the prohibitions of the Valar, through the Shadowy Seas and past the Enchanted Isles, till Eärendil reaches Aman and walks up from the coast towards Valinor, the Guarded Land. There he is hailed by a messenger as:

'the looked for that cometh at unawares, the longed for that cometh beyond hope! ... bearer of light before the Sun and Moon ... star in the darkness, jewel in the sunset, radiant in the morning!'

(*The Silmarillion*, chapter 24)

The nearest parallel to language of this sort, with its -eth endings and its Biblical phrases, is in the psalm-like announcement of the eagle to Gondor in *The Return of the King* (see p. 209 above); and like the eagle's announcement this one carries significant ambiguity. The ambiguity had probably been present from Tolkien's first explorations.

Tolkien had been struck by the name, or the word *Earendel* as early as 1914, when he encountered it in an Old English poem, now titled (rather unimaginatively) *Christ I*. The lines go, *Eala earendel...*, 'O Earendel, brightest of angels, sent to men above Middle-earth', but it is not clear even from context what *Earendel* means or even whether it is a proper name. Tolkien would however soon have realized, from obvious sources like the standard edition of the poem and Grimm's *Teutonic Mythology*, that (as with 'Light-elves' and 'Dark-elves') the material for an imaginative reconstruction was ready to hand, and one which would once again 'save the evidence'. In the first place the Old English poem (apart from the word *earendel*) is a translation of a known Latin antiphon, which begins *O Oriens* ... This antiphon is however, in both Latin and Old English, one of a series representing the cries of the patriarchs and prophets still in Hell, before the coming of Christ to release them, calling out for a Saviour for those who 'sit in darkness and the shadow of death', or a prophet who will announce the coming of the Saviour; the *O Oriens* one is taken to refer to John the Baptist. But whatever *earendel* is, the image that goes with it is people in sorrow looking up from the darkness and hoping both for rescue and for light. Despite this strongly Christian context, though, *Earendel*, if it is a name, also has pagan connections. Aurvandil (the Old Norse equivalent of Old English Earendel) is present in Snorri Sturluson's *Prose Edda* as a companion of the god Thor. The two went together on an expedition, but as the god waded across the freezing rivers of Élivágar he had to put his weaker companion in a basket. Aurvandil's toe stuck out, got frostbite, and was broken off and thrown into the sky to become a star. Mythographers have been consistently puzzled by this allusion – one of them suggested that since his wife was called Gróa, 'to grow', Aurvandil might represent the seed-corn which in Scandinavia is sometimes sown too early and killed by the frost. However, what Tolkien might have taken from it is, first, confirmation that Earendel/Aurvandil is the name of a star, and second, that it was a sign of hope and good tidings to pagans as well as to Christians.

All this provides a suggestive background for Tolkien's tale of Eärendil, and a justification for the Biblical language of the herald's announcement quoted above. In Tolkien the people looking up from darkness and seeing a great light are not the patriarchs and prophets of Old Testament story, but the inhabitants of Middle-earth; and the great light they see is not Christ coming to harrow Hell and release them, but the Silmaril announcing the rescue mission of the Valar. The setting is indeed not Christian but pagan, or at least pre-Christian. However, if pagans knew of Aurvandil, and Aurvandil is the same linguistically as Earendel, and Earendel was early equated with Christ, then could pre-Christians not have had some intuition, some sense of a forerunner of their true and eventual Saviour? In *Mere Christianity* C.S. Lewis calls these 'good dreams . . . queer stories scattered all through the heathen religions' about a god who brings men 'new life'. Tolkien would probably not have approved Lewis's phrasing, nor did he mean in any way to confuse Eärendil or the Valar with Ilúvatar, the Creator. But in a clearly limited and deliberately imperfect way, Tolkien's *Silmarillion* closes with an analogue of intercession, forgiveness, and salvation coming necessarily from

outside a ruined Middle-earth, just as it began with analogues of
the Fall of Angels and the Fall of Man.

To return to the questions implied on p. 239 above, does it
not seem presumptuous to repeat Christian myth in a work of
human imagining, and what is the point of doing so, with variations? Not surprisingly, the answer seems to be the same as that
given already for *The Lord of the Rings* (see pp. 180–2 above).
Tolkien wanted all his life to bring together the Christian religion
in which he devoutly believed, and the relics of the pre-Christian
beliefs of his ancestors embedded in the literature which he spent
his professional life studying. He had no sentimental feelings
about paganism or heathenism, which he hated (see the discussion of Denethor above, pp. 176–7); but he was not prepared
to write off everything pre-Christian as irrelevant, unlike his
countryman Alcuin (see the discussion of Frodo above, pp. 183–
5). His re-interpretation of Eärendil is accordingly not presumptuous, but respectful. And as for the point of the variations on
Christian myth, one might say this.

Tolkien knew that 'angel' meant originally *angelos*, 'messenger'.
But there could be several kinds of messenger. Gandalf is one –
very unlike the traditional image of an angel, with his long beard
and short temper, but an 'angel' just the same. Eärendil is a
second, announcing the coming of the Valar to Morgoth and
Middle-earth alike. Galadriel, in a way, could be seen as a third.
Of course she is not a Maia, like Gandalf, and she also bears
some share of the responsibility for the Fall of the Noldor and
the exile from Middle-earth, joining in the rebellion against the
Valar because of her yearning 'to see the wide unguarded lands
[of Middle-earth] and to rule there a realm at her own will' (*The
Silmarillion* chapter 9, but see the alternative versions discussed
in *Unfinished Tales*). If, then, Galadriel were to be remembered
at some later date as equivalent in status to Gandalf, and so an
'angel', she would have to be a *fallen angel*; and if fallen angels
are the same as devils, then this seems inconceivable. Fallen angels

are not however the same as devils in all opinions and all traditions. In some traditions, including early English ones, some of the angels exiled from Heaven with Satan became devils, but others, more undecided or more neutral, became elves. At Judgement Day some of these may regain forgiveness and salvation and return to their old home, as Galadriel does at the very end of *The Return of the King*. This still does not make Galadriel into an angel, even in the sense of a messenger, in the way that Gandalf is; but one can imagine how a human being, looking back at the events of the Third Age and the First Age 'from some historical distance', as suggested above, could be confused, could put together Galadriel the Noldo exiled by the Valar and Gandalf the Maia sent by the Valar (both of them allowed in the end to return) and no longer be able to see much difference.

Tolkien knew further that the Greek word for the New Testament, *euangelion*, contained within it the *-angel-* element, and meant 'the Good Message', neatly translated into Old English as the *gód spell*, the 'good story', the Gospel. In modern English 'spell' no longer means 'story', but 'enchantment', but Tolkien might have thought that example of semantic change entirely appropriate and perhaps not even accidental. Gospel means Christian message; means good story; means powerful enchantment. Angel means winged creature of Christian myth; means messenger; means elf. Earendel means John the Baptist, the forerunner of Christ the Saviour; means star; means seed – though the 'seed that does not die and cannot be destroyed' in the very last sentence of the 1977 *Silmarillion* is the seed of evil which will 'bear dark fruit even unto the latest days', i.e. till now, and beyond now. These complexes of meaning suggest that history, and linguistic change, keep on generating new meanings for words and demanding new versions of story, even when they are the same words and the same story. In that case *The Silmarillion*, centred as it is on the sins of possession and mastery and the desire to exercise skill whatever the consequences, becomes less a mythology for England and more one for its own time, for the twentieth century: a myth re-told, with proper respect for what in myth is unchanging, because myths always need retelling.

Some comparisons

For all that has been said, *The Silmarillion* can never be anything other than hard to read. And for all Stanley Unwin's tact, back in 1937, it is unlikely that it would ever have been published in any form, let alone so many of its forms, without the prior success of *The Lord of the Rings*. It has no hobbits – those essential mediating figures which provide the modern audience with a focus and a point of relationship. It scorns novelistic convention: in the 'Narn i Hîn Húrin' Tolkien began to develop the process of detail, of the verisimilitude created by subordinate characters and extraneous dialogue which the modern reader expects, but it did not go very far. Even towards the end of his writing career and after forty or more years of development he was clearly still unsure how to bring in some features, like the 'Dragon-helm of Dor-lómin' or the animated 'Black Sword' itself. It is not that such motifs could not be used in a modern environment – one can imagine one of Tolkien's many emulators in the field of fantasy integrating them into a commercially successful fantasy – rather, that Tolkien continued to reach (every bit as much as James Joyce) for something beyond the conventionally publishable.

Like Joyce with *Finnegans Wake*, he demanded too much for most audiences. Christopher Tolkien has stated that 'To read *The Silmarillion* one must place oneself imaginatively at the time of the ending of the Third Age – within Middle-earth, looking back', (*The Book of Lost Tales I*, p. 4), which is certainly correct. It parallels Tolkien's own image of *Beowulf* as 'a poem from a pregnant moment of poise, looking back into the pit, by a man learned in old tales who was struggling, as it were, to get a general view of them all' (*Essays*); as it does Lord Macaulay's determined attempt to see Livy or Virgil as men looking back on the old lays or ballads of their own tradition, and seeing in those lays or ballads men who were looking still further back, to 'the brave days of old'. If Tolkien had a literary model *for The Silmarillion*, furthermore, it must surely be the *Prose Edda* of Snorri Sturluson, a compendium of pagan mythological materials made by a man who was himself not at all a pagan, but who did not wish to see the old traditions of the poetry of his native tongue vanish for ever; a man who also continually illustrated his prose synopses with quotations from poems, too often poems which have otherwise vanished. The scholarly attraction of reading a work like that is hard to convey, for one feels at once the interest of the material that is there, regret for all the material that is not there (but which clearly could have been if only the author had thought it worth while), and the constant stimulus to the imagination of the gaps and omissions and lacunae. 'Heard melodies are sweet', says the poet, 'But those unheard are sweeter', and in cases like Snorri's one might almost believe it to be literally true. Tolkien corroborated the thought with the remark (in a letter to Christopher in January 1945) that 'A story must be told or there'll be no story ... yet it is the untold stories that are most moving' (see *Letters*, p. 110, and also *The Book of Lost Tales I*, p. 3). One can agree and sympathize. Just the same, appreciation of this kind is rare, recondite, and hard to develop: an acquired taste *par excellence* – though one can see it as strikingly exemplified as anywhere in Christopher Tolkien's own 'introduction' to another late synopsis preserving intensely moving and suggestive scraps of older tradition, *The Saga of King Heidrek the Wise*.

Christopher Tolkien, the editor of all the material discussed here, has also suggested that one's model for how to respond should be Sam Gamgee's innocent and naïve response to Gimli's song of Moria in *The Fellowship of the Ring*, where 'great names out of the ancient world [Nargothrond and Gondolin] appear utterly remote': 'I like that! ... I should like to learn it' (*The Book of Lost Tales I*, p. 3). This is so. But there is an alternative response to ancient story in *The Two Towers*, on 'The Stairs of Cirith Ungol', when Sam has just given yet one more version of the tale of Beren and Lúthien, and remarked that he and Frodo still appear to be in the same tale: perhaps some hobbit-child in the future will demand the story of 'Frodo and the Ring'. Yes, says Frodo, and he will demand 'Samwise the stouthearted' too: 'I want to hear more about Sam, dad. Why didn't they put in more of his talk, dad? It makes me laugh.' This does not count for much as literary criticism, but it does make a point, which is that *The Silmarillion*, very much unlike *The Lord of the Rings*, stays resolutely on the level of 'high mimesis' or above, eschewing humour, detail, fine texture. It is able conversely to appeal to qualities virtually ruled out in even the most ambitious commercial fantasies: stoicism, nonchalance, irony, magniloquence. But here it has not been followed, and probably cannot be. *The Silmarillion* is most likely to be seen – paradoxically, for things were meant to be the other way round – as a further and immensely detailed 'Appendix' to *The Lord of the Rings*. The 'lost lore that was used in the making of a new thing' is, in this case alone, no longer lost. It tells a great deal about the 'making', but it also returns attention to the 'new thing'.

CHAPTER VI

⠂⇒◉⇐⠂

SHORTER WORKS:
DOUBTS, FEARS, AUTOBIOGRAPHIES

Tolkien's shorter works considered

No one could call Tolkien *homo unius libri*, a one-book man –
Hammond and Anderson's *Descriptive Bibliography* of 1993 lists
twenty-nine 'Books by J.R.R. Tolkien', thirty-six more 'Books
Edited, Translated, or with Contributions' by him, and thirty-nine
'Contributions to Periodicals', taking 349 pages to do it; and the
list has been subsequently extended. However, these totals include
a good deal of repetition and reprinting, and a substantial amount
of posthumous publication, while many, indeed most of the items
listed are brief or 'occasional': if it had not been for the later
celebrity of their author they would be completely forgotten. If
one considers that Tolkien was a professional academic, in a trade
devoted to publication and in a position intended to free one for
it (for few British academics manage thirty-five years in successive
university Chairs), then one has to concede that by normal standards he did not publish very much – apart, of course, from *The
Hobbit* and *The Lord of the Rings*, which if not 'one book' may
at least be considered one related sequence.

If one excludes posthumous publications – which include,
besides *The Silmarillion* and *The History of Middle-earth*, the three

children's books, *The Father Christmas Letters* (published in 1976), *Mr Bliss* (1982), and *Roverandom* (1998) – the remainder may be divided into: academic works; poems; and short narrative pieces. The academic works are considered briefly below, but one may say even more briefly here that apart from the collaborative 1925 edition of *Sir Gawain and the Green Knight* (which probably got him his first Oxford Chair), and the admittedly ground-breaking and field-defining essay on *Beowulf* eleven years later, Tolkien published less academically than most of his colleagues, and especially little after about 1940. From an early period, though, he had kept on publishing individual poems, sometimes pseudonymously, and nearly always in relatively obscure locations – college and university magazines, small or privately-printed collections. A list of these is given in Humphrey Carpenter's *Biography*, and extended in Hammond and Anderson's *Bibliography*. The count comes to about thirty, though the exact total may vary depending on how one deals with Tolkien's habit of reprinting poems in different places, but also rewriting them more or less extensively. The count further excludes both his contributions to the 1936 collection (again privately-printed and little-circulated) *Songs for the Philologists*, and the sixteen poems in the 1962 collection *The Adventures of Tom Bombadil*, nearly all of them reprints or rewritings, as also the many poems in *The Hobbit* and *The Lord of the Rings* (themselves in their turn often rewritings).

One is left with a small handful of narrative pieces which Tolkien wrote and published in his own lifetime: 'Leaf by Niggle' (1945), *Farmer Giles of Ham* (1949), and *Smith of Wootton Major* (1967). To these one might add the verse-dialogue section of 'The Homecoming of Beorhtnoth Beorhthelm's Son' (1953), which is in itself fiction but is framed by academic commentary, and so especially hard to classify. Of these four pieces I would regard 'Leaf by Niggle' and *Smith of Wootton Major* as quite clearly autobiographical allegories, in which Tolkien commented more or less openly on his own intentions, feelings and career; while both *Farmer Giles of Ham* and 'The Homecoming of Beorhtnoth Beorhthelm's Son', disparate though they are, also tell us something about the tension in Tolkien's mind between the demands of his job and his increasingly urgent drive towards non-academic creation. In their different ways these pieces, backed up by the early poems and some of the posthumously-published fragments, take us closer to Tolkien's inner life than do his major works.

Autobiographical allegory: 1
'Leaf by Niggle'

Tolkien seems to have published 'Leaf by Niggle' almost as a whim. He was asked on 6th September 1944 by the editor of *The Dublin Review* (we do not know why) for a story which would help his magazine to be 'an effective expression of Catholic humanity', and sent him 'Leaf by Niggle' on 12th October. By Tolkien's standards this was practically by return of post, and would not have happened if the work had not already been in existence, probably for some time. In his *Biography* Humphrey Carpenter suggests that it was written close to the date of submission, and was born of Tolkien's 'despair' at his failure to finish *The Lord of the Rings*; but it seems more likely that it had been written some five years earlier, just before the outbreak of World War II, though it probably did arise out of the author's 'preoccupation' with *The Lord of the Rings* (*Letters*, p. 257). Tolkien himself said, in the 'Introductory Note' to *Tree and Leaf* (1964), that he woke one morning with the whole thing 'already in mind', and that it took him 'only a few hours to get down, and then copy out'. One of its sources, according to the 'Note', was grief or anger at the fate of a tree, a 'great-limbed poplar tree' which 'was suddenly lopped and mutilated by its owner, I do not know why'. However, it is not hard to see other and more personal sources for the anxiety, and the defiance, which Carpenter quite correctly senses in the work. Though it has a general bearing on 'Catholic humanity' as a whole, and so fits the original editor's commission rather well, it is at once a personal apologia, and a self-critique: expressed, as often in Tolkien (for all his prot-estations, see p. 161 above) in the form of strict or 'just' allegory.

Allegorical meaning is signalled at once by the first sentence: 'There was once a little man, called Niggle, who had a long journey to make.' The reason for his journey is never explained, nor how he knows that he has to make it. But there should be no doubt as to what this means. The Old English poem *Bede's Death-Song* begins, in its original Northumbrian dialect, *Fore thaem neidfaerae*, '(Be)fore the need-fare'. A 'need-fare', or 'need-faring', is a compulsory journey, a journey you have to take, and that journey, Bede declares, begins on one's *deothdaege* or 'death-day'. So the long journey the 'little man' Niggle has to make – which all men have to make – is death. The image is at once 'as old as the hills', completely contemporary, and totally familiar. This is the easiest of the equations in the extended allegory.

But if everyone has to take it, why is the central character of the story called 'Niggle', and not, for instance, Everyman? Here the *OED* gives a highly relevant definition. 'To niggle', according to the *OED*, means 'To work . . . in a trifling, fiddling, or ineffec-tive way . . . to work or spend time unnecessarily on petty details; to be over-elaborate in minor points'. This was certainly a vice of which Tolkien could be accused. One can see it in some of his posthumous publications, like the 1982 edition of *Finn and Hengest*, edited from Tolkien's notes by Alan Bliss. Considering that this was only Tolkien's second publication on *Beowulf*, and that his first has remained the most influential and frequently-cited publication on the poem of all time, it might seem amazing that it has had no academic impact at all – no one ever cites it. But it is extremely hard to follow, detail-crammed past ready comprehension. Many of Christopher Tolkien's notes on sections of *The History of Middle-earth* create a similar impression, of his father constantly working and re-working on minor, or as the *OED* calls them, 'petty' details, with the result that nothing at all (except *The Hobbit* and *The Lord of the Rings*) ever got finished. Much of this work was essential for the success of *The Lord of the Rings*, as has I hope been demonstrated in chapters II to IV above. But much of it was work wasted. Tolkien singles out 'niggling' here and there as a vice in his fiction – Sador in the 'Narn î Hin Húrin' is a niggler who 'spends much time on trifles unbidden', see *Unfinished Tales*, p. 64. It is a failing of which Tolkien was conscious, and which he ascribed to himself ('I am a natural niggler, alas!' he declared in a letter to Rayner Unwin in 1961). Though the 'long journey [Niggle] had to make' is death, still, Niggle should be equated not with Everyman but with Tolkien.

Niggle is a painter, and Tolkien of course was (pre-eminently) a writer. Though his paintings have proved surprisingly attractive, many of them collected in the 1979 volume *Pictures by J.R.R. Tolkien*, they were nearly always illustrations of his writing. Turning back to Niggle, how good a painter is he? This is one of the two main questions raised in paragraphs two to five of the story, and the answer is fairly complex. He is certainly 'Not a very successful one', partly because 'He was the sort of painter who can paint leaves better than trees'. But the 'one picture in particular which bothered him', while it started as 'a leaf caught in the wind', soon became a 'tree', indeed a 'Tree', while behind it 'a country began to open out; and there were glimpses of a forest marching over the land, and of mountains tipped with snow'. If one translates this from Niggle to Tolkien, it makes good sense. Tolkien began with short poems like the 1914 'Lay of Earendel' (leaves, so to speak); they grew into explanatory narratives (like the unpublished 'Book of Lost Tales' or the 'Quenta Silmarillion'); as he wrote on a 'country began to open out' (see the account of the early stages of *The Lord of the Rings*, pp. 60–65 above);

and there were indeed glimpses in it of a 'forest marching' (the Ents). When it says that 'Niggle lost interest in his other pictures; or else he took them and tacked them on to the edges of his great picture', one might relate this to decisions like Tolkien's introduction of 'The Adventures of Tom Bombadil', the 1934 poem, to *The Fellowship of the Ring*, see again p. 62 above. In an earlier work I suggested that the 'leaf' was *The Hobbit* and the 'Tree' *The Lord of the Rings*, but this has been rendered doubtful by the continuing appearance of unpublished works which show how relatively late those two works developed. One would have to say now, more vaguely, that Niggle's 'great picture', existing only in the mind, was something like a completely finished and integrated version of the whole history of Arda from Creation to the end of the Third Age.

But the other reason for Niggle's lack of success, besides trying to paint things 'too large and ambitious for his skill', is (and this is the reason mentioned first) that 'he had many other things to do'. In particular he has a house, a garden, many visitors, and an annoying neighbour, Mr Parish. What, one has to wonder, if one accepts that this story is in fact an allegory, and therefore dependent on the making of equations (see pp. 162–4 above), are these supposed to represent? A good deal fits together if one remembers the particular circumstances of Tolkien's job. He was a Professor, and an Oxford Professor. The capital letter is significant, for (as is not the case in the USA), not all faculty members at Oxford University were or are Professors: indeed few of them are, and even fewer in Tolkien's time. A Professor is the holder of a University Chair, and in Tolkien's time the English Faculty at Oxford had precisely three of them, the Rawlinson and Bosworth Chair of Anglo-Saxon which Tolkien held from 1925 to 1945, and the two Merton Chairs, by convention one for literature and one for language, the latter of which Tolkien held from 1945 till his retirement in 1959. These three Chairs are valuable and (it is fair to say) coveted by the very much larger number of college fellows and university lecturers – in Tolkien's day some thirty or forty – who compete for them. (One might note that C.S. Lewis, despite the distinction of his scholarship, was one of the thirty or forty who never got one at Oxford, moving to Cambridge to take one up in 1954 when he was fifty-five.) The Chairs are valuable primarily because they are University appointments, not college appointments, and release their holders from the very time-consuming task of responsibility for providing undergraduate tutorials in colleges (in my day, some twelve to sixteen teaching hours a week). In return for this relative freedom Chairs are required to give a certain number of open lectures for undergraduates (Carpenter's *Biography* says thirty-six a year, which I think should be thirty-five, five sets of seven), to teach graduate students (there were relatively few of these in Tolkien's time), but above all to advance scholarship in their subject-areas, primarily by publication.

Tolkien may well have felt increasingly uneasy about this. By 1939, when 'Leaf by Niggle' was written, he had had two major academic 'hits', as said above, in the 1925 *Gawain* edition and the 1936 *Beowulf* lecture – and that is two more than most academics ever manage, and bears comparison with most Professors. But they were not being followed up: the projected *Pearl* edition appeared without his name on it (see p. 197 above), and the only sequel to the *Beowulf* lecture was the posthumous *Finn and Hengest* edition (and in a way the edition of the Old English poem *Exodus*, also appearing only posthumously, in 1981). Furthermore Tolkien had made a very considerable mark on Middle English studies with a 1929 article on the dialect of a group of early texts from Herefordshire (another of his favourite West Midland counties). He was certainly expected to follow this up with a major book-length study, or sequence of studies, but did not do so. His edition of one of the texts, *Ancrene Wisse*, did appear in 1962, after he had retired, but it is only a transcript: no notes, no glossary, and an introduction written by another scholar. Tolkien could have framed a defence against accusations of misuse of his time and favoured Professorial position. Carpenter's *Biography* points out that in his second year as Professor, Tolkien gave not thirty-six 'lectures and classes' but 'one hundred and thirty-six', though that is still not a heavy teaching load. More significantly, Tolkien had several students and collaborators like Mary Salu and Simone D'Ardenne who were extremely grateful to him and who continued his work, especially on the Herefordshire texts. Just the same, the accusations were certainly being made, if not in 1940, then before many years had passed. One hears an echo of them in J.I.M. Stewart's Oxford-based sequence of *romans à clef*, 'A Staircase in Surrey', in the third of which, *A Memorial Service*, we come upon two people discussing 'Professor Timbermill':

> 'A sad case,' [the Regius Professor] concluded unexpectedly.
> 'Timbermill's, you mean?'
> 'Yes, indeed. A notable scholar, it seems. Unchallenged in his field. But he ran off the rails somehow, and produced a long mad book – a kind of apocalyptic romance.'

'Timbermill', of Linton Road, is a philology professor, author of *The Magic Quest*. The connection with Tolkien, the philology professor from Northmoor Road, is obvious, and the 'ran off the rails' verdict is one that was often passed by Oxford insiders.

All this is reflected in the ominous sentence early on in 'Leaf by Niggle', 'Some of [Niggle's] visitors hinted that his garden was rather neglected, and that he might get a visit from an Inspector'. In Tolkien's day there were no 'Inspectors' at British universities – there are now, in the shape of the five-yearly Research Assessment Exercise – and one cannot tell quite what these hints in reality amounted to. They certainly applied to Tolkien, though. In a letter of 1958 he apologized for delay in reply, saying that he had been on leave, given so that he could:

complete some of the learned works *neglected* [my emphasis] during my preoccupation with unprofessional trifles (such as *The Lord of the Rings*): I record the tone of many of my colleagues.

(*Letters*, p. 278)

With all these data in mind, one can then continue the interpretation of the allegory, which becomes both more personal and more touching. The 'many other things' which Tolkien, like Niggle, had to do could be seen as his lectures, graduate supervisions, faculty meetings, examination setting and marking, involvement in faculty appointments, etc. They may not seem to amount to very much in total time, by normal office standards, but as any Professor will confirm, they break up almost every working and writing day, to the detriment of concentration. Niggle seeming 'polite enough' when visitors come to call, but 'fiddl[ing] with the pencils on his desk', may be an image of Tolkien trying to listen to his colleagues and his students but 'thinking all the time about his big canvas' (or his fiction-writing), out 'in the tall shed that had been built for it out in his garden (on a plot where once he had grown potatoes)'. One might think of the shed as Tolkien's own study, eventually in a converted garage; but it, and the garden, and indeed the potatoes, bear a less literal meaning.

The 'neglected' garden, for a start, which causes the hints about the Inspector, surely represents the professional field which Tolkien had been appointed to cultivate, and which some certainly thought, see above, he had indeed been neglecting. Niggle once grew potatoes in this garden; and Tolkien once wrote the academic articles expected in his field (about a dozen of them by 1939, but almost none in normal form thereafter). Niggle now uses the garden only as the place for a shed in which to do his painting; and Tolkien certainly exploited all his academic knowledge in his fiction, as indicated many times in the chapters above. But of course the visitors do not see the painting. In any case some of them come because they want his 'pleasant little house', know that he will eventually have to vacate it, are busy calculating when, and wonder 'who would take [it], and if the garden would be better kept'. Ingrown academic societies are continually abuzz with rumours about Chairs being vacated, by death or retirement, and with speculations about who will get, or who deserves, the next appointment. Tolkien can hardly have avoided knowing this, and mentions junior colleagues yearning to take over his 'padded seat'. What he wanted, like Niggle, was no doubt for someone in authority to tell him to forget everything else, get on with what he really wanted to do, and award him a 'public pension' to do so. But while an Oxford Chair might look like a sinecure from outside, Tolkien knew that it was by no means *sine cura*, 'without a care'; he commented wryly in his letter that the 'padded seat' was actually 'stuffed with thistle'. As for the public pension, that was a pipe-dream. What was needed was 'some concentration, some work, hard uninterrupted work, to finish the picture, even at its present size' (and the last phrase, this time, might really be taken to refer to *The Lord of the Rings* alone).

The scene-setting above covers only the first three or four pages of 'Leaf by Niggle'. The story's action begins, in the 'autumn' (Tolkien was by this time in his late forties), with 'a knock on the door'. The knock is Niggle's neighbour, Parish, and it has to be said that he is not so easy to fit neatly into the allegory being developed. He is extremely annoying. He has no respect for Niggle, often criticizes the state of his garden, and has no interest at all in his painting, which he sees as nothing but a waste of time and resources. He has no conscience about interrupting Niggle's work (which he does not see as work), and sends him off, at a time when Niggle feels, correctly, that he has no time to spare, to fetch a doctor for his wife, and a builder for his wind-damaged house. This trip in the wet prevents Niggle from ever finishing his picture, and gives him the chill from which he has only just begun to recover when first the long-dreaded

Inspector calls, and then the Driver appears to take Niggle on his 'journey'. One might say that Parish kills Niggle, but this is not quite strictly true; what he does is kill his time. And there are a few things to say for Parish, and they are said, by the narrator or (in the end) by Niggle. Parish's wife, it turns out, did *not* need a doctor, but Parish *was* genuinely lame, in pain, and without a bicycle. In times gone by (and this is what Niggle says in the Workhouse after he has gone on his journey), 'He was a very good neighbour, and let me have excellent potatoes, very cheap which saved me a lot of time'. The successful or 'eucatastrophic' end of the story depends on Niggle and Parish co-operating, so much so that 'Niggle's Picture' and 'Parish's Garden' combine, to become 'Niggle's Parish'.

Tolkien wrote in 1962 that the name 'Parish' was just 'convenient, for the Porter's joke', at the same time denying that 'Leaf by Niggle' was an allegory at all – he preferred to see it as 'mythical' – because Niggle was 'a real mixed-quality person, and not an "allegory" of any single vice or virtue'. But allegory does not have to renounce mixed qualities. It is attractive to see Niggle and Parish as a 'bifurcation', as two aspects of Tolkien's own personality which he wished he could combine: the one creative, irresponsible, without ties (Niggle is not married, but Parish is), the other scholarly, earthbound, practical, immediately productive (preoccupied, one might say, with the duties of his limited 'parish'). The speculation may seem more plausible if one notes that there is a much more evident bifurcation in the near-simultaneous appearance of the Inspector of Houses, and the Driver who comes to take Niggle on his journey: 'Very much like the Inspector he was, *almost his double*: tall, dressed all in black' (my emphasis). And there is a third in the Two Voices whom Niggle hears discussing his case at the end of his time in the Workhouse, the severe First Voice, the gentle Second Voice. All three pairs seem to represent, in their different ways, Tolkien's own mixed judgement on Tolkien.

At the centre of this is a severe judgement on 'niggling'. Niggle goes on his journey (dies). Although he had not entirely forgotten that he was going to do so, and had begun 'to pack a few things in an ineffectual way' (to make spiritual preparations), this amounts to little, for his bag contains neither food nor clothes, only paint and sketches, and in any case he loses it on the train (for some things cannot be taken with you). He is removed to the 'Workhouse', which is clearly Purgatory. And there he learns to stop niggling, in the *OED* sense given above. 'He had to work hard, at stated hours', at jobs which are useful but of no interest. What he learns is to manage his time:

> he could take up a task the moment one bell rang, and lay it aside promptly the moment the next one went, all tidy and ready to be continued at the right time. He got through quite a lot in a day, now; he finished small things off neatly.

This, of course, is what he needed to learn in life, and it brings him, not 'pleasure' but at least 'satisfaction'. He becomes 'master of his time' and loses the continuous 'sense of rush'. Many academics and office-workers will sympathize with his problem and wish they could reach the same solution. But what has it cost?

In the world which Niggle has had to leave, it appears to have cost everything. The Inspector of Houses had the last word, and he agreed with Parish. Niggle's picture was of no value. He should have used the resources it represents to fix Parish's house. 'That is the law.' Public opinion strongly agrees with him. The narrator had said early on that though the Tree was 'curious. Quite unique in its way', 'I dare say it was not really a very good picture', while Niggle, though also unique, was at the same time 'a very ordinary and rather silly little man'. The last phrase is repeated without all the qualifications in the story's penultimate scene, when Councillor Tompkins says 'he was a silly little man'. He is contradicted by Atkins (all the characters in this scene, NB, have

'Huggins'-names, not 'Baggins'-names, see p. 8 above), but Atkins is 'nobody of importance, just a schoolmaster'. It is Tompkins who gets Niggle's 'pleasant little house' in the end. All Niggle's canvases are used as patching, and though Atkins manages to preserve a fragment of one, and even to frame it and have it hung in the Town Museum, in the end both Museum and leaf are burned and Niggle and all his work 'were entirely forgotten in his own country'. His earthly epitaph is (from Perkins), 'poor little Niggle! ... Never knew he painted'. This fate of oblivion must have been what Tolkien feared; in 1939, and 1944, it seemed all too likely. The story's title, 'Leaf by Niggle', is ironically enough the same as the title of the fragmentary work first preserved in the Museum and then lost for ever. Tolkien on other occasions at about this time (see *Sauron Defeated*, pp. 303, 308) gloomily imagined his writings surviving only uncomprehended and unread. In 1944 it might well have seemed to Tolkien that the story 'Leaf by Niggle' would indeed (apart from *The Hobbit*) be all that would ever be left of thirty years of writing.

However, that is only half of the story. As at the climax of *The Lord of the Rings* (see pp. 210–11 above), 'Leaf by Niggle' offers what is this time a *narrative* bifurcation. The real world, the live world, dismisses and forgets Niggle: from that point of view, Niggle's story is a tragedy. The *other* real world, the world after death, turns to 'eucatastrophe'. The Two Voices discuss his case, and the Second Voice (Mercy, I would suggest, as opposed to Justice, two of the four Daughters of God), makes the statement for the defence. Niggle *could* paint; but he was humble; he did a good deal of his duty; he expected nothing for it, not even grati- tude; in the end he sacrificed himself in something like full aware- ness of what he was doing. Tolkien might have hoped that the first four points at least would also apply to him. Niggle's reward is to find his picture come true at the end of his journey, his 'sub-creation' accepted by the Creator, there in full detail and 'finished' but (exactly the opposite of what happens in the world he has left) not 'finished with', in the sense that there is still enormous scope for development. But for the development he needs, and gets, Parish, whose time in the Workhouse has clearly been spent in making him *less* practical, not more so. In the end Niggle is ready to turn his back even on the Tree and go on elsewhere, under the guidance of a man who 'looked like a shepherd' (with obvious Christian suggestion). The Great Tree and the country round it are left, however, in the heavenly country, as 'a holiday', 'a refreshment', and even an 'introduction to the Mountains' for others – which by a further irony, which Tolkien must have deeply appreciated, is what they have become, in their way, in our real world, through the publication of so much of his 'picture', finished and even unfinished, but by no means finished with. 'Leaf by Niggle' ends as a comedy, even a 'divine comedy', on more levels than one. But while it looks forward to 'divine comedy' it incorporates and springs from a sense of earthly tragedy: failure, anxiety, and frustration.

Poems written and rewritten

These feelings are to some extent mirrored, even increasingly mirrored, in Tolkien's thirty or so separately-published poems. Some are entirely comic, like the two poems on 'Bimble Town' (a British seaside resort), 'Progress in Bimble Town' and 'The Dragon's Visit'. Others are exercises in difficult metre, like 'Errantry' (its long development into 'Bilbo's Song at Rivendell' is explained in *The Treason of Isengard*). Quite a few could be described as 'Missing Link poems', or 'ancestor-poems', poems which Tolkien wrote, like Macaulay's *Lays*, to fill a gap in literary history. Examples include Tolkien's 1923 'egg' riddle in Anglo-Saxon, his two 'nursery-rhyme' poems from the same year (see pp. 25–26 above), and the two Old English 'trapped mortal' poems from *Songs for the Philologists* reprinted and translated in

Appendix B to *The Road to Middle-earth*. These latter could again readily be seen as reconstructed 'ancestors' for much later works like Keats's 'La Belle Dame Sans Merci'. Conversely, in 1923 Tolkien picked out a single line from *Beowulf, iumonna gold galdre bewunden*, 'the gold of men of old, enmeshed in spell', and constructed a whole poem with the line as title. The poem deals with the way that the spell of treasure betrays and corrupts its successive owners, elf, dwarf, dragon and man, a theme Tolkien returned to in *The Hobbit* and later, see pp. 48–9, 169–70 above. He rewrote and republished this poem with the same title in 1937, and as 'The Hoard' in 1962 and 1970. It and some of the best of his comic or light-hearted poems were picked out and reprinted in *The Adventures of Tom Bombadil* in 1962, while others remain unpublished. According to Hammond and Anderson's *Bibliography* there were six 'Bimble' poems. The most extensive of Tolkien's 'Missing Link' poems, also still unpublished, may well be 'The New Lay of Sigurð' (*Sigurðarkviða hin nyja*), said to have been written, entirely characteristically, to fill the gap in the hero Sigurð's adventures caused by the loss of a number of pages from the one major surviving manuscript of the *Elder Edda* poems.

From very early on, however, Tolkien wrote and published a number of poems which are not so light-hearted, which indeed seem to be preoccupied with the themes of mortality and immortality. Some of them, but not all, are related to the developing 'Silmarillion'. Thus the poem which he published as 'The City of the Gods' in 1923 is given a context within Tolkien's mythical geography as 'Kor' in *The Book of Lost Tales I*); the same is true of 'The Happy Mariners' (1920, and again 1923, this time with an Old English title, '*Tha Eadigan Sælidan*', later in *The Book of Lost Tales II*); of 'The Lonely Isle' (1924, but see *Letters*, p. 437); and 'The Nameless Land' (1927, but contextualized as 'The Song of Ælfwine' in *The Lost Road*). Nevertheless these poems, all rather like each other, say something even as first published, detached as they then were from the 'Silmarillion'.

They are in the first place static: visions, or perhaps one should say observations, of a city, a country, an island. 'The City of the Gods', a sonnet, simply describes a stone city, without inhabitants, without noise of any kind, standing under hot afternoon sun. 'The Lonely Isle', two twelve-line stanzas plus a single-line coda, has in it fairies dancing and a bell pealing, but once again no narrative movement, not even vestigial story. 'The Nameless Land', sixty lines in the extraordinarily difficult *Pearl*-stanza form, starts off with the word 'There', and again describes a Paradisal landscape by the sea, inhabited only by dancers who cannot be human for there 'no man may be'. 'The Happy Mariners' has slightly more movement in it, and begins 'I know a window in a western tower'. Looking out of the window, the observer sees 'fairy boats go by . . . through the shadows and the dangerous seas, / Past sunless lands to fairy leas', these latter being no doubt the 'nameless land' where 'no man may be'. But the observer in the tower cannot follow them; they are 'happy', he is not. 'The Nameless Land' ends with an image of longing, 'The lights of longing flare and die', as does 'The Lonely Isle', 'I long for thee and thy fair citadel', but the longing has no power: the last coda-line of 'The Lonely Isle' is 'O lonely, sparkling isle, farewell'. Within the geography of the 'Silmarillion', all this can of course be fitted together, as indeed it is in 'The Song of Ælfwine' – Ælfwine is an Anglo-Saxon sailor who in one version reaches, in others has a vision of Tol Eressëa, the 'nameless land' now named. But as the poems stand, the impression they make is of an observer in this world haunted by visions of an ideal landscape which he cannot reach, a Paradise from which he is for ever barred. The observer is mortal, the land is for immortals.

In a second, later group of poems, Tolkien imagined a mortal getting closer to immortality. These poems, though, are sadder than the visions, and became even sadder as they were rewritten. The first of them, like 'Leaf by Niggle', was presumably sent in reply to a request, this time from the Sisters of the Sacred Heart

Convent at Roehampton, and was published in their *Chronicle* in 1934. In thirteen eight-line stanzas it tells the story of a girl, Firiel, who goes out from her parents' house at dawn and sees an elf-boat going by. The elves invite her to join them. She asks where are they going – 'To Northern isles grey and frore'? No, they reply, they are going 'to Elvenhome / beyond the last mountains', to the bell, the tower and the sea-foam of Tolkien's earlier visions; few, they say, receive this offer to leave the mortal world where 'grass fades and leaves fall'. Firiel takes one step towards the boat, but then her heart fails her and the elves leave. She goes back in the house, 'Under roof and dark door', the dew and the vision fading. She is re-absorbed into the mundane world: housework, talk of 'this and that', but also breakfast. The poem's last words are 'please, pass the honey'. Tolkien rewrote this poem as the last in the collection *The Adventures of Tom Bombadil*, where it has become 'The Last Ship'. Once again Fíriel (as she is now) goes out, sees the elf-boat, receives the invitation, asks the question, is given the answer. This time, though, it is made clear that this is the *last* boat, there is room for only *one* more, this is 'our *last* call' (my emphasis). Fíriel stops this time not because 'her heart misgave and shrank' but because 'deep in clay her feet sank'. She calls out to the elves, 'I cannot come ... I was born Earth's daughter'. She goes back this time not only 'under roof and dark door' but also 'under house-shadow'. And, most significantly, the two-stanza scene of busy chatter and cheerful breakfast has been removed. She no longer dresses in 'green and white' but plain 'russet brown', and goes simply 'to her work'. All that is said of the morning is 'Soon the sunlight faded'. The poem ends with a single stanza saying that the world is still there but there is no more passage out of it – a stanza which ends, like the one before it, with the word 'faded'. In 'The Last Ship' the sense of loss, and of death, is very much stronger than in 'Fíriel', and is presented as an inevitable loss. Fíriel is made of 'clay', like all children of Adam, she is 'Earth's daughter', and this is the fate she has to accept – indeed her name, we are told in the 1962 mock-editorial 'Preface', simply means 'mortal maiden'. Hers is an 'anti-fairy story', about the 'Escape from Death' rejected.

Tolkien rewrote another 1934 poem, 'Looney', in a very similar way in *The Adventures of Tom Bombadil*, as 'The Sea-Bell'. Both versions consider what happens to humans who do manage to reach Elvenhome, and who by old tradition (as in Keats's 'Belle Dame Sans Merci') are never the same again. The rewriting is however in this case even more extensive – 60 lines in 1934 become 120 in 1962 – and the editorial 'Preface' adds further ominous suggestion. In 'Looney' the speaker who narrates the main portion of the poem (the 'looney' or lunatic of the title) is asked at the start, 'Where have you been; what have you seen / Walking in rags down the street?' In reply he says he comes from a land where he met no one. He got there by sitting in an empty boat, which took him unbidden 'to another land' – seemingly, the land of flowers and bells and unseen dancers of the 'vision' poems above. 'Looney', unlike Fíriel or the longing visionaries, has then actually reached the 'nameless land'. But something in it goes wrong, with the appearance of a 'dark cloud'; the spring vanishes, the leaves fall, the sea freezes. He gets back in the empty boat and comes home, to find that all the 'Pearls and crystals' he gathered in the strange country have become 'pebbles', the 'flowers' 'withering leaves'. All he has left is a shell, in which he hears an echo, perhaps of music – the poem does not say. It is traditional, of course, for 'fairy gold' to turn to leaves, and for the people the elves reject to be unable to return to society, often because everyone they once knew is dead, see p. 89 above. The 1934 poem can be understood within this convention.

The 1962 poem, however, no longer has the query at the start, or the identification of the speaker as a 'looney', while in the 'Preface' to *The Adventures of Tom Bombadil* Tolkien, in his role as editor of a hobbit-manuscript, remarks that it is 'certainly of hobbit origin', is 'the latest piece' in the collection, and 'belongs to the Fourth Age', i.e. the age which begins after the War of the Ring is over. It is included in a collection of poems from the Third Age, though, 'because a hand has scrawled at its head *Frodos Dreme*'. One wonders what is the effect of this last piece of Tolkienian invention. Is it not yet another 'bifurcation'? At the end of *The Lord of the Rings* Bilbo and Frodo, unlike Ælfwine or Fíriel or any of the other visionaries, do leave Middle-earth and make the Great Escape to 'white shores and beyond them a far green country'. Only very shortly before, though, Frodo has been quite sure that 'I am wounded . . . it will never really heal'. As with the earthly and heavenly endings of 'Leaf by Niggle', or with Sam and Frodo lying down in Mordor to die, and being rescued in their sleep by the eagles, we are given in a way two endings, the 'eucatastrophic' one which is validated by the story, and a tragic one which seems equally if not more plausible: 'It's the way things are, Sam'. 'The Sea-Bell', or 'Frodos Dreme', gives the normal ending for those, humans or hobbits, who follow visions of immortality, not the exceptional one set aside for Ring-bearers. They hear the 'sea-bell' ringing, they get in the empty boat, they find themselves transported to the glimmering jewel-land with its unseen dancers and far-off music which haunted Tolkien's imagination. As in 'Looney', something goes wrong. But in 'The Sea-Bell' it seems to come as a punishment for hubris. The speaker makes himself a mantle, a wand, a flag and a crown, and names himself 'king of this land', calling on the inhabitants to show themselves. Immediately the cloud comes, the country is disenchanted, becomes a land of beetles, spiders and puffballs, the speaker realizes he is old, 'years were heavy upon my back'. The boat brings him back, he has nothing, and even the shell is now 'silent and dead', without even an echo.

> Never will my ear that bell hear,
> never my feet that shore tread.

Not only is the speaker exiled; his vision has gone too. In 'The Sea-Bell' one may see Tolkien turning his back on the very notion of the Great Escape, and on images which had been with him for close on fifty years.

The Lost Road

The most persistent of those images, as has been shown above, is that of the land far across the sea, Westernesse, the Blessed Land, the Earthly Paradise, the Land of the Undying. This was not just Tolkien's image, for hints of such a belief are scattered widely across North-European literature. King Arthur is taken away across the sea to be healed of his wounds and to return one day from Avalon; in Tolkien, Avallónë is a place on the Lonely Isle, Tol Eressëa. The legend of the Drowning of Atlantis was known already to Plato; Atalantë is the Quenya or Elf-latin word for 'the Drowning of Númenor'. There are versions in several languages of the story of 'the Navigation of Saint Brendan', which Tolkien was to rework as 'Imram', the most successful of his later 'poems of (im)mortality' (see pp. 288–9 below). Less known than any of the above, and still unexplained, there is the mysterious opening of Beowulf, which tells how a child is sent across the sea to become King Sceaf, or King Sheave, of the Danes, and is eventually sent back in a boat after his death to 'those who sent him' (not, note, 'Him who sent him', though the Beowulf-poet was certainly a Christian). The belief in a Land of the Dead, or of future life, across the ocean must be the reason, Tolkien argued, for the Anglo-Scandinavian custom of burying kings and nobles in their ships, or under ovals of standing stones arranged to look like ships. It is hard, indeed, to think of a better explanation.

As so often, Tolkien adapted all these fragmentary suggestions for his own uses, and tried at least twice, in his centrally creative years, to make them into an extended and publishable story.

These attempts survive as 'The Lost Road', written probably in 1936 and published in 1987 under the same title; and 'The Notion Club Papers', written probably in 1944 and appearing in 1992 as part of the volume *Sauron Defeated*. In framing both of these, however, Tolkien faced an obvious problem: the discovery of America. Saint Brendan or medieval Arthurians might have been able to believe in a supernatural land across the ocean to the West, but this is no longer credible. The problem was solved, in Tolkien's developed if never quite perfected mythology, by the idea of 'the Lost Straight Road' – a good further example, incidentally, of the use of myth to mediate between incompatible beliefs (see p. 179 above). According to Tolkien, Aman, the land of the Valar, was cut off *twice* from Middle-earth. The first time comes after the defection of the Noldor and the return of Fëanor and his followers to Middle-earth. After this the Valar fill the seas between Aman and Middle-earth 'with shadows and bewilderment'. There is no physical separation but just the same, 'the Blessed Realm was shut' (*The Silmarillion*, chapter 11). The way is opened again by Eärendil, piloted by the Silmaril, and the human allies of the Valar are rewarded not only with the new land of Númenor, raised from the ocean, but with new sight of the coasts of the Undying Land. In the end, though, corrupted by Sauron, and by resentment and fear of death, the king of Númenor sails to invade Aman with an armada, and to gain immortality by force. The Valar, so to speak the archangels of Earth, lay down their government of Arda and call upon the One, Ilúvatar, God the Creator. And he 'changed the fashion of the world'. Númenor and its armada are drowned; Aman and Eressëa are removed for ever from the world, 'into the realm of hidden things'; 'new lands and new seas' are made in their place – this must be the creation of America; and (though Tolkien does not say this quite openly) the world is for the first time made round. When at the end of the 'Akallabêth' section in *The Silmarillion* the descendants of the surviving Númenóreans put to sea again to look for the land in the West which they know was there, they find only lands 'subject to death', and those who press on furthest find themselves in the end back where they started, so that they say 'all roads are now bent'. Their loremasters continue to insist, though, that there must still be a Straight Road, for those permitted to find it, which leads off the Earth and through the atmosphere and through space, 'as it were a mighty bridge invisible', leading to the Lonely Isle and even beyond to Valinor. It must be this road which is taken by Frodo and company at the end of *The Lord of the Rings*; this road which Fíriel is invited to take, and (perhaps) which 'Looney' has taken. There may be others, Tolkien suggests, who 'by some grace or fate or favour of the Valar' find the hidden gateway. As the hobbits sing, probably without understanding what the words mean:

> Still round the corner there may wait
> A new road or a secret gate . . .

The 'problem of America' continued to bother Tolkien into old age. He remained dissatisfied with the solution given above, wondering if such a myth was still acceptable in a scientific age, see *Morgoth's Ring* pp. 369–83, and Hammond's article in *Tolkien's 'Legendarium'*. His doubts might have been assuaged if he had remembered that there was distinguished precedent for ambiguity, in *Paradise Lost* X 668–91, where Milton carefully wrote in a passage about the Earth's tilt allowing his readers to believe either a traditional/mythical or a modern/scientific explanation. A more pressing problem – it was the one Tolkien continually faced – was that of converting image into story. John Rateliff has suggested, in another article in *Tolkien's 'Legendarium'*, that the famous decision of C.S. Lewis and Tolkien (recorded most clearly in *Letters* p. 378) to write a story each, the one about space-travel, the other about time-travel, was triggered by Lewis's reading of Charles Williams and realization that it was possible to write such a thing as a 'philosophical thriller'. Lewis then produced *Out of the Silent Planet* (1938), with the two further volumes of the 'Space trilogy' to follow, the third of them, *That Hideous Strength* (1945) making evident if mis-spelled cross-reference to Tolkien. Tolkien, however, found himself in a blind alley.

It is fairly clear what he meant to do in 'The Lost Road'. As with word-sets like 'dwarf'-*dvergr*-*Zwerg* (see p. xiv above), he was struck by the consistency and continuity of names in Germanic tradition, and felt it must mean something. He knew that the names Alboin and Audoin, found in a famous story in Peter the Deacon's *History of the Lombards*, were just different spellings of Old English Ælfwine and Eadwine, or indeed modern Alwyn, Edwin. He knew also that they meant respectively 'elf-friend' and 'bliss-friend', while there was a third name in Old English, Oswine, which meant 'god-friend', or at least 'friend of the pagan gods', the *osas*, the Old Norse *Æsir*. Names like Oswald and Oswine had been borne, however, by Christians and indeed by saints, and Tolkien was prepared, rather daringly, to identify the *osas/Æsir* not with demons, but with the demi-gods or archangels or Valar of his own mythology. His idea was to do as he had done with the opening chapters of *The Hobbit*, and put traditional story into a new and this time a contemporary, rather than just an anachronistic context. The names would continually recur. There would be a story about three modern English characters, named rather confusingly Oswin, Alboin and Audoin; another one about an Anglo-Saxon called Ælfwine, which would draw in the legend of 'King Sheave' from *Beowulf*; one about the famous Lombard pair (this would have been hard to manage, given its ferocity and triumphal skull-goblet); and one set in Númenor before its fall, where the names would have become Valendil, Elendil, Herendil, again respectively 'friend of the Valar, elf-friend, bliss-friend'. The main point of the story as a whole, presumably, would have been the recurrence of the theme of loyalty to an ideal in different settings. The difficulty was that the same story repeated, whatever the settings, was bound to be monotonous. Mr Rateliff argues that Tolkien was getting there, and that the story might have succeeded as a 'philosophical thriller' if Tolkien's attention and his energies had not been drawn off by the many problems connected with getting *The Hobbit* into print, in 1937, but the contrast with Lewis's companion-piece is not encouraging. Five thousand words into *Out of the Silent Planet*, its hero has been kidnapped and is on a space-ship heading for Mars on a mission of conquest. Five thousand words into 'The Lost Road' and the characters are still considering the history of languages, the story as yet invisible.

Amazingly, however, Tolkien tried again, and at a much less propitious time. In 1944 Tolkien was two-thirds done with *The Lord of the Rings* and might have been expected to be pressing on to its conclusion – he had, after all, a publisher ready, waiting, and keenly supportive. Instead he spent a very considerable amount of time and effort, which he concealed perhaps slightly guiltily from Stanley Unwin (see *Sauron Defeated*, p. 145) on a second 'Lost Road' effort, 'The Notion Club Papers'. This moves even more slowly than 'The Lost Road', with an assortment of characters loosely based on the Inklings discussing dreams, languages, the works of C.S. Lewis, and time-travel. The characters include at least two Tolkien self-images in Ramer (the name means, I think, 'Looney', see once again *Sauron Defeated*, p. 150) and Rashbold (a translation of 'Tolkien'). Several of the same elements as in 'The Lost Road' recur, especially the continuity of names – one character is Alwin Arundel Lowdham, i.e. Ælfwine Éarendel. Visions of Númenor and of Anglo-Saxon England appear, along with a poetic version of 'King Sheave', and this time 'The Navigation of Saint Brendan' is brought in, as a poem which was to be published in 1955 as 'Imram'. At the centre of both versions lies the sentence (in Old or Proto-Germanic, a non-existent or asterisk-language), *Westra lage wegas rehtas, nu isti sa wraithas*, 'the Straight Road lay west, now it is bent': one might note the ominous adjective *wraithas*, 'bent'. Tolkien also repeatedly quoted some lines from the genuine Old English poem 'The Seafarer' about the longing to go to sea, which he expanded with lines of his own invention, so as to convert the sea-longing into longing to *cross* the sea and reach the *wlitescéne land,/eardge-ard ælfa and ésa bliss*, '[a land lovely to look on], the dwelling place of the Elves and the bliss of the Gods'. The last line neatly brings together Tolkien's three ruling name-elements, Ælf-, Os-, Ead-, 'elf-, god-, bliss'.

There are, one may concede, some moments of pathos in Tolkien's two 'lost road' inventions. When in the earlier of them Alboin reads to his father Oswin the lines from 'The Seafarer' which say, no one knows 'what longing is his whom old age cutteth off from return (*eftsíth*)', his father looks at him and says the old know perfectly well that they cannot return, and that while they are cut off from *eftsíth*, they are not cut off from – he uses the word *forthsíth*, going away, going forth, but he means death (*The Lost Road*, p. 44). That is the journey he has to take, like Niggle – or like Fíriel, a name which re-appears without explanation in *Lost Road*. In 'The Notion Club Papers' Tolkien built in a suggestion that the Oswin-equivalent had found the Straight Road in his boat Éarendel (unless it hit a mine), but this is a withdrawal (*Sauron Defeated*, p. 234). The two 'lost road' stories which Tolkien tried so hard to write, in 1936 and 1944, in tandem with *The Hobbit* and *The Lord of the Rings*, centre like 'Firiel' and 'Looney' on the fear and the acceptance of mortality, the vision and the rejection of the Earthly Paradise. But the theme could not be integrated into adventure. Its most finished and perfect expression is the 1955 poem 'Imram' (reprinted in *Sauron Defeated*) in whose 132 lines Saint Brendan tells a younger man of his journeys: he saw a mountain (the remains of Númenor), an island and a Tree (perhaps Tol Eressëa), and a Star 'where the round world plunges steeply down / but on the old road goes' (the Silmaril at the end of the Lost Straight Road). But he does not take the Straight Road, the Star is now only 'In my mind', and even Saint Brendan accepts the common fate:

> Saint Brendan had come to his life's end
> under a rain-clad sky,
> journeying whence no ship returns;
> and his bones in Ireland lie.

Popular lays, a lai, and an anti-lay

There is no hint of any of the themes above in *Farmer Giles of Ham*, Tolkien's only entirely successful published narrative outside the hobbit-sequence. It seems to have been conceived at about the same time as *The Hobbit*, when the Tolkien children were still small, and to have reached its eventual published shape by 1938 (see *Bibliography*, pp. 73–6). I have remarked above (p. 58) on its root in the interpretation of place-names. I have also remarked elsewhere (in *The Road to Middle-earth*) on its potential as an allegory, and sketched out such a reading. I freely concede, however, that this is probably *furor allegoricus*, or allegorist's mania: *Farmer Giles of Ham* makes too much sense as a narrative in its own right to need an allegorical reading, and is furthermore entirely light-hearted. Some of its cheerfulness may come from the fact that it is set neither in real history nor in the world of Tolkien's mythology, but in the entirely spurious history, long accepted but long since disproved, of the 'Brutus books' of *Sir Gawain*, of *King Lear*, and indeed of the 'Old King Cole' of nursery-rhyme, all of which are referred to. It is very much in a 'Never-Never Land', and Tolkien felt no urge to take any of it seriously – so that we have in it the overpowering figure of Farmer Giles (a kind of anti-Beowulf, with his extremely amateurish preparations for fighting the dragon), several of the best human-dragon conversations in literature to set against Bilbo and Smaug, and a whole cast of comic minor characters, from Garm the dog to Augustus Bonifacius Ambrosius Aurelianus Antoninus, the proud tyrant of the Little Kingdom, not to mention the miller, the giant, Sunny Sam the blacksmith, the grey mare and the parson-grammarian.

Yet entirely in line with the light-heartedness, the story makes a point, and a rather aggressive one. In the mock-editorial 'Foreword', a device Tolkien liked, we are told that the story is just like the histories erected on Macaulay's hypothesized *Lays*, i.e. not contemporary with the events it records, 'evidently a late compilation', and 'derived not from sober annals, but from the popular lays to which its author frequently refers'. As he does: popular lays are mentioned at least five times in the story. After his accidental victory over the giant Giles finds himself 'the Hero of the Countryside', and comes home from market 'singing old heroic songs'. These may well be the same as the 'popular romances' which give Farmer Giles's sword its name Tailbiter; and it is certainly the 'tales about Bellomarius' the dragon-slayer which give Giles the heart to face the dragon. The knights are singing a 'lay' – an old one, from before the tournament-era they live in – when the dragon surprises them; they should obviously have paid more attention to it. And after Giles's defiance of the king, 'it was impossible to suppress all the lays which celebrated his deeds', and indeed it is this which prevents the king from raising an army. All these songs and lays and romances, of course, have disappeared before the time of the mock-editor's 'Foreword', as is usually the case with lays of this type. But they are associated in the story with Giles; with the 'vulgar tongue'; with 'plain heavy swords' like Tailbiter, made to do their job; with being out of fashion. By contrast the king's court is associated with magniloquence, book-Latin, style at the expense of substance (the Mock Dragon's Tail made by the confectioner), and a reluctance to take old tales seriously. There is no doubt about which of these two sides wins, and one could say that the conflict between them is more serious than the one between Giles and Chrysophylax, for in the end the two latter form an alliance.

One could also say that the situation at the centre of *Farmer Giles of Ham* is rather similar to the situation in Laketown when Bilbo arrives there in chapter 10 of *The Hobbit*. The ruling class or official Establishment has no time for 'old heroic songs' or the creatures mentioned in them, being entirely concerned with money; the king's knights regard dragons as mythical (and of course vice-versa). There are also those who do remember old tradition, but have got entirely the wrong idea from it; they are the people in *The Hobbit* who expect the river to start running gold immediately, or the people in Ham who crowd round Giles calling him 'Hero of the Countryside' and praising 'The Glory of the Yeomanry, Backbone of the Country' (etc.). In between are figures like Giles and Brand, at once traditional and practical. All these had their counterparts in the twentieth society Tolkien lived in, and it is clear that he was mocking at once the 'Correct and sober taste' of his literary contemporaries (see *Essays*, p. 16), and something like the taste for easy fantasy which has become more widespread since his time.

However, the only person in the story who gets things right nearly all the time, besides Giles (and the grey mare), is the parson of Ham. He is treated with a certain comedy, especially in his first recorded conversation with Giles. He insists on seeing the sword presented by the king (one may wonder why? What makes him suspect something?). He looks carefully at the letters on scabbard and blade, but 'could not make head or tail of them'. He covers up with bluff, of a highly professional kind – 'The characters are archaic and the language barbaric'. Nevertheless, he gets the answer right in the end: the sword is Caudimordax, or Tailbiter. After Giles's first victory over Chrysophylax, the parson gets things wrong in his rather pompous speech setting the terms for Chrysophylax's compensation-payments. Or does he? The dragon has no intention of keeping the many oaths he swears, and though this may have been:

beyond the comprehension of the simple, at the least the parson with his booklearning might have guessed it. Maybe he did. He was a grammarian, and could doubtless see further into the future than others.

This does not make sense in modern terms, in which grammar has nothing to do with foresight. But it does in medieval terms, in which grammar was the same as 'glamour' (the ability to change one's shape and deceive observers) and as 'grammarye' (magic). The parson is at any rate the first to realize the dragon is not coming back; and, critically, he is the one who tells Giles to take some stout rope with him, 'for you may need it, unless my fore-sight deceives me'. It does not, the rope enabling Giles to bring back both dragon and treasure, so that the parson deserves his eventual bishopric and handsome rewards. In all this one might see a for once self-flattering image of Tolkien's own trade: runology paying off, linguistic skills abetting plain horse sense – with, of course, the business-like skills of a self-consciously modern world gratifyingly and correspondingly humiliated.

Farmer Giles of Ham, unlike 'Leaf by Niggle' and the poems discussed above, shows Tolkien quite at ease with himself. In this it remains the exception in his minor works. An anxiety which may have weighed with him was his own increasingly detailed and powerful work of 'mediation' (see pp. 179–82 above) between the Christian world of, for instance, the parson-grammarian, and the pagan traditions of older mythology (which are not allowed into *Farmer Giles* at all). One could put an orthodox case against him like this. Tolkien was presenting his Valar as benevolent archangels subject to the One, the Creator, but he showed a certain tendency towards identifying them with the pagan deities of his ancestors – Tulkas, for instance, Tolkien's warlike Vala, is very like Snorri's image of the god Tyr, and his name could easily be seen as a Proto-Germanic word for 'warrior', *tulkaz*. Tolkien was by no means the first Christian philologist to veer in this direction, and a case could be made out in his defence. C.S. Lewis's *Mere Christianity* suggestion about the 'good dreams' of heathen religions has already been mentioned; and Tolkien seems to accept something like this through the 'splintered light' thesis embedded in his poem 'Mythopoeia', quoted in Carpenter's *Biography*. One *could* make out the case, but it was by no means the view of the Catholic Church, and runs directly contrary to the view of Alcuin (see p. 181 above). Tolkien's famous and much-quoted assertion, to a Jesuit friend, that '*The Lord of the Rings* is of course a fundamentally religious and Catholic work' has an air of defensiveness about it. How far could a believing Christian go in dealings with pre- or non-Christians? This question animates both the long poem 'The Lay of Aotrou and Itroun' (published 1947) and the verse dialogue 'The Homecoming of Beorhtnoth Beorhthelm's Son' (published 1953).

The word 'lay' in the title of 'Aotrou and Itroun' means something different from all previous uses in this book. There are in Old French a number of poems called *lais*, usually 'Breton *lais*', pre-eminently the twelve *lais* of Marie de France. Some eight Middle English poems also call themselves 'Breton lays' in imitation. Tolkien's 'Aotrou and Itroun' closely follows the form and matter of these, using 'Britain/Brittany' and 'Briton/Breton' interchangeably. It is derived (as Jessica Yates has shown in her essay of 1993) from a late Breton ballad about a 'Corrigan', a witch, or fairy, or shape-shifter with malevolent powers. What seems original to Tolkien is the poem's stern morality. In his version, the story begins with a noble Breton couple, Aotrou and Itroun, who remain childless. Instead of bearing their fate humbly, Aotrou, the lord, calls on the Corrigan for a fertility potion, not agreeing the fee. The potion works, Itroun bears twins, but the Corrigan's fee (she has now turned from hag to beauty) is a demand for Aotrou's love. He refuses, is cursed, and dies, as does Itroun, of grief. The refusal, the curse, and the death of Aotrou are there in the Breton ballad; but in Tolkien alone the death is deserved, or at least prompted by Aotrou's attempt to sway Providence by supernatural forces. Tolkien's moral is clear and unequivocal. Aotrou's sin lay not in submitting to the Corrigan – he defied her as successfully as Sir Gawain warded off the Green Knight's lady, and with a firm profession of Christian faith – it lay in having any dealings with her at all. The poem ends:

> God keep us all in hope and prayer
> from evil rede and from despair,
> by waters blest of Christendom
> to dwell . . .

This complete rejection of supernatural allure goes several stages beyond the ambiguity of poems like 'Looney'.

The 1953 essay-cum-poem on 'The Homecoming of Beorht-noth Beorhthelm's Son' is harder to summarize, but briefly again, it consists of three sections, a short foreword which sets the historical scene – a battle the English fought and lost against Viking raiders in Essex in August 991, commemorated in the Old English poem *The Battle of Maldon* – and explaining what follows; 'The Homecoming' itself, a dialogue in alliterative verse between two characters, Torhthelm and Tídwald, come to retrieve the headless body of their master Beorhtnoth from the battlefield, seen as a coda to *The Battle of Maldon*; and a short section on 'Ofermod', which provides an academic critique of the Old English poem. This latter has been highly influential. Its six pages firmly reject the view of Maldon put forward by previous scholars, including Tolkien's old colleague and collaborator E.V. Gordon, who had edited the poem in 1938, and W.P. Ker, who had called it 'the only purely heroic poem extant in Old English', and the best example of a native 'heroic lay' that we have. Tolkien argued that Gordon, Ker, and the rest, were completely wrong. The poem is not a celebration of the heroic spirit but a deep critique of it and of the rash and irresponsible attitudes it created. It was the alderman Beorhtnoth's *ofermod* (Tolkien translates 'overmaster-ing pride') in giving the pagan Vikings a sporting chance which led to his own death and the deaths of his men. He had no right to make such a gesture, and the poem should be read as making this point.

Tolkien's opinion seems to be challenged in the original Old English poem, if not outright denied, by the famous speech of the old retainer Beorhtwold, who says as he refuses to retreat (Tolkien's translation again):

> 'Heart shall be bolder, harder be purpose,
> more proud the spirit as our power lessens!'

Beorhtwold at least offers no criticism of his leader. Tolkien deals with this problem in the verse-dialogue – and this is entirely personal, quite without scholarly warrant – by taking the lines away from their context on the battlefield, putting them into a dream, and adding to them (and this is not translation but original composition):

> 'Mind shall not falter nor mood waver,
> *though doom shall come and dark conquer*'
> (my emphasis).

The lines have become clearly pagan, indeed Manichaean, and the older and more sensible speaker in the dialogue, Tídwald, immediately identifies them as such: 'your words were queer ... It sounded fey and fell-hearted / and heathenish, too: I don't hold with that'. Tolkien has a thumb firmly in the balance here, even more than he does elsewhere in the dialogue, though this is one-sided enough. Torhthelm (the name means 'Shining Helmet') is a young fool whose mind has been addled by the heroic lays he repeatedly mentions, while Tídwald ('Time-Rule, Time-Ruler') is a wise veteran who understands what the lays really mean: 'you can hear the tears though the harp's twanging'. As with 'Aotrou and Itroun', the point is made very evident. There should be no compromise. Christian men, even Christian warriors, should not be seduced by the 'theory of courage' which Tolkien had himself praised seventeen years before (see *Essays* p. 20), or by 'the doctrine of uttermost endurance in the service of indomitable will', as expressed by Beorhtwold. This was exactly the 'Ragnarök spirit' which had led to and been painfully crushed in World War II. Heroic lays, then, were entirely misleading – or at least, needed very careful revision amounting to rewriting to make them safe.

'Aotrou and Itroun' was published in 1947, *Farmer Giles of Ham* in 1949, 'Beorhtnoth' in 1953. We know, however, that *Farmer Giles* was in existence in substantially its present form in 1938, while a fragment of an early version of 'Beorhtnoth' survives from even earlier (see, respectively, *Bibliography* pp. 73–4 and *The Treason of Isengard* pp. 106–7). It is tempting to see 'Aotrou' and 'Beorthnoth', the *lai* and the anti-lay, as signs not exactly of the darkening of Tolkien's imagination (it had always had its dark side, though this increases significantly between 'Firiel', say, and 'The Last Ship'): rather, of a certain self-doubt. Was it acceptable to rewrite myths? How would this be seen by the Two Voices in the Workhouse?

Autobiographical allegory: 2 *Smith of Wootton Major*

The genesis of *Smith of Wootton Major* is more easily traced than that of any of Tolkien's other fictions. In 1964 Tolkien was asked to write a Preface to a new edition of *The Golden Key*, a work by the nineteenth-century writer of fantasy and fairy-story, George MacDonald (a major figure in C.S. Lewis's 1946 allegory of death and judgement, *The Great Divorce*). He agreed, and started work in January 1965 – when he was 73 – intending at this stage to tell a story of a cook and a cake by way of explaining the meaning of the word 'fairy', or 'Faerie'. But the story grew and the Preface was abandoned, the story being published eventually in 1967 as *Smith of Wootton Major.*

In my view this last of Tolkien's completed fictions demands to be read in very much the same way, if with more difficulty, as 'Leaf by Niggle', as what I have called an 'autobiographical allegory'. The view has been strongly resisted, for instance by David Doughan, in an article of 1991, and by Verlyn Flieger, who gives a long account of *Smith of Wootton Major* in her 1997 book *A Question of Time.* Flieger has further had the advantage of reading an unpublished essay on the story by Tolkien himself, in which he glosses the meaning of his own work; though Flieger reports that in the essay Tolkien appears to be 'conducting a running argument with himself on the question of whether the story is or is not an allegory', and veering from plain statement that it is not to reluctant concession that some of it evidently is. In particular, Tolkien states that 'The great Hall is evidently an "allegory" of the village church', while 'the Master Cook . . . is plainly the Parson and the priesthood'. Nevertheless, one remark which Doughan, Flieger and Tolkien himself all cite with great approval is Roger Lancelyn Green's comment in a review of the story that it should not be allegorized, for 'To seek for the meaning is to cut open the ball in search of its bounce'.

Green's comment is of course itself a short allegory, of exactly the kind which Tolkien himself deployed tactically in his *Beowulf* lecture, i.e. a *reductio ad absurdum.* Cutting open a tennis-ball to look for the bounce is evidently absurd, so cutting open a story must be the same. If, that is, a story is like a tennis-ball. I would suggest that the analogy is mistaken. An allegorical story ('Leaf by Niggle', say) is much more like a crossword-puzzle. In both cases, solutions which are too easy are no fun. In both cases, however, it helps to have a few easier solutions to start with (like realizing the 'journey' Niggle has to take is death). The reader then knows that the harder solutions you go on to look for must not be incompatible with the more obvious ones; and the more correct solutions that are filled in, the narrower becomes the range of possibilities for what is left. The strong point of my analogy is that the attraction of both allegories and crossword-puzzles is an intellectual one – the reader of an allegory is actively engaged, not being passively led. The weak point is that solving a puzzle gives only a fleeting feeling of satisfaction; but solving an allegory gives in addition a new awareness which entirely reshapes one's understanding of the surface narrative. Further-more, at the more advanced stages of reading an allegory, it is not essential to come up with the one single correct solution (which, again unlike crossword-puzzles, may never have existed). A suggestive or a provocative one will do. The story continues to make sense, or in the terms of Green's allegory to 'bounce', even after it has been 'cut into'. Indeed it bounces better, it makes more sense. Only the laziest of readers make no effort at all to respond to the clues given by authors of allegory.

The main clue by which one detects the presence of allegory is always something strange or inadequate in the surface narrative, such as the unexplained nature of Niggle's certain but unmotiv-ated journey. Situations like that do not arise in real life; so, this is not quite 'real life'. This seems to me to be the case with *Smith*. Wootton Major is a perfectly plausible name for an English village, but the village itself is strange. It is centred on a Great Hall, 'built of good stone and good oak' but 'no longer painted or gilded as it had been once upon a time'. Furthermore the village seems to centre on cooking, on the office of Master Cook, and on institutionalized feasts, such as the Twenty-four Feast, held only every twenty-fourth year, for which the Master Cook must bake a Great Cake. None of this has any resemblance to any English custom one can recognize. The strangeness provokes one to ask whether it all means something else; and Tolkien supplied the answer (wrote in a guide solution, so to speak) with his remark that Hall and Cook were allegories of church and parson. In which case, one has to remark, Wootton Major is in a fairly degraded state, the spiritual functions of the Church (as Tolkien also commented) 'steadily decaying . . . into mere eating and drinking – the last trace of anything "other" being left in the children'.

Not that there is more than a trace left in the children. The most spiritually degraded figure in the story is the second in the succession of (four) Master Cooks which it contains, the stop-gap Nokes. Nokes is not much of a Cook, is furthermore sly, derivative, and malicious, but most damning of all, and strongly insisted on, he has very little idea of fantasy or of Faerie. It might be better if he had none (the Fairy Queen tolerantly disagrees), but unfortunately he does still have a memory of the concepts. He associates fantasy with childishness; accordingly creates a childish image of it in his sugar-icing Great Cake, which will be 'pretty and fairy-like' and have a 'Fairy Queen' doll in the middle of it; and shows it to the children, who then respond exactly on cue (a few of them), clapping their hands and crying, 'Isn't it pretty and fairylike!' Nokes here surely represents many of the things which Tolkien most disliked, and which he presented in different ways from the Master of Laketown in *The Hobbit* to the 'misologists' of his 'Valedictory Address to the University of Oxford', those 'professional persons' who 'suppose their [own] dullness and ignorance to be a human norm, the measure of what is good' (see p. xvi above). The last clause describes Nokes very accurately. He has only a weak, childish, pink-icing notion of fantasy himself, but assumes that that is all there can ever be; and since he is well aware of the feebleness of his own imagination, he assumes all images of the fantastic, of Faerie, must be feeble too. One has to say that this seems to be the problem with many of Tolkien's professional critics.

In Wootton Major, then, spirituality has dwindled down to

materialism, and authority-figures are liable to lead children in exactly the wrong direction, and against their natural inclination. What forces are operating against this? The major ones are Smith himself, Alf Prentice, and the 'fay-star' which passes from person to person. The fay-star is the story's central symbol. Smith is present as a nine-year-old at the story's first Twenty-four Feast, when he swallows the star without knowing it – it had been put in the cake by Nokes, with the agreement of Alf Prentice, though Nokes has no idea of what it may do, and Alf does. For most of his life this star marks Smith out as different from his fellow-villagers, and acts as his passport into Faery; it enables him to find the entrance unknown to others. However, unlike Looney or Firiel or anyone else in Tolkien's works, Smith acquires the power to go into Faery and return freely, while when he does return he is not crazy and unsocial – like Looney or the hobbits we hear of taken by the 'wandering-madness' – but a particularly valuable and admired member of the village community. He makes things that are 'plain and useful', but others 'for delight'. He seems to have reached in life the balance which characters like Niggle and Parish could only achieve jointly, and then only after death. The star seems, then, to represent something like Tolkien's own impulse towards fantasy, the quality of vision; while Smith represents the ideal response to it, using it as an enrichment of normal life rather than a distraction. In this view the story begins to look like another 'mediation', this time a successful one, between fantasy and reality.

Yet Smith's situation is not ideal. There is a strong stress in the story on succession and inheritance. The star came in the first place, we learn, from Smith's maternal grandfather Rider, once the Master Cook, also a visitor to Faery, but someone who in the end leaves and does not come back – someone who makes the Great Escape of Tolkien's imagination. Smith inherits the star from him, but he does not inherit Rider's position. Also, unlike his grandfather, when he has to return the star he loses his passport into Faerie and any chance he had of making the Escape: to use Oswin Errol's terms, he is still condemned to *forthsíth*, but knows there can be no *eftsíth*. To make the blow harder to bear – Tolkien uses the phrase 'a great weariness and bereavement' – Smith knows the star will be passed on (indeed he decides that it shall be passed on) not to his own blood, though he has a son, Ned, to whom he is deeply attached, but to the great-grandson of the stupid and hostile Nokes. The one succession, of Master Cooks, entirely passes Smith by: it runs Rider – Nokes – Alf Prentice – Harper. The other succession, of owners of the star, includes him only temporarily: it runs Rider – Smith – Nokes's Tim. Smith's own bloodline, Smith – Nan – Ned, is passed over both for authority and for vision. Smith lives a good life, but it ends in multiple disappointment. He can never return to Faery; the star is passed on; he cannot give it to Ned. It is true that there is a coda to the story, in which Alf Prentice reveals himself as the King of Faery and squashes Nokes's pretensions in satisfying style, rather like Farmer Giles declaring independence from Augustus Bonifacius; true also that Smith ends content with the new Cook, Harper, the Hall restored to its former glory, and his wife's nephew inheriting the fay-star. But even then Nokes has the last word.

One explanation for this mixture of success and failure, happiness and bereavement, may be that we are looking once more at one of Tolkien's narrative 'bifurcations', with elements of his own life projected into both Smith and Alf Prentice. Smith is very different from Niggle in his all-round real-world competence, his continuous making of useful things – though one might remember that Tolkien always insisted that he too had spent his life doing useful things (teaching, supervising, examining) for other people, and that the activity had been fostered, not as many people said 'neglected', as a result of his excursions into fantasy. It is easy to see Smith as a Tolkien-figure. And he is very like Niggle, after all, in his dominating visions of a world elsewhere.

Alf Prentice provides an authority and a ratification which is quite absent from 'Leaf by Niggle'. His appearance on Smith's side (and very much against Nokes) is rather as if, in Tolkien's earlier allegory, some figure had appeared in Niggle's world to reprove Councillor Tompkins thunderously and have Niggle's pictures displayed with suitable respect. Furthermore, if one remembers the Cook/parson equation made by Tolkien himself, the revelation that a Cook can also come from Faery and entirely approve of fay-stars very much supports Tolkien's deep belief (or desire to believe) that his gift for fantasy in no way compromised his religion. Not only are fantasy and reality harmonized in *Smith*, one might conclude: so also are fantasy and faith.

There is one further point to be made about *Smith* before concluding. Its structure is clear, if unusual. After a few pages in which Wootton Major and its customs are introduced, the first extensive scene is the one in which Smith receives the star, at the first Twenty-four Feast. The next extensive scene in which main characters converse is two Feasts later, forty-eight years, when Alf is getting ready to bake his second Great Cake and wants the star back to put in it. Smith's life has gone by in between. And most of the space in between deals with his visions of Faery: the elf-warriors returning from their ships, the King's Tree rising into the sky, the Wind of the World which strips the weeping birch of its leaves, the dance where he receives the Living Flower, finally the scene where he recognizes the Queen of Faery and is returned to 'bereavement'. It is very hard to say what these visions mean, if anything; they may just be meant as examples, like the stages of St Brendan's 'Imram'. The birch, however, certainly had particular and personal symbolic meaning for Tolkien. It stood for philology. It stood for the 'B-scheme' of education which he introduced to Leeds, and tried to introduce to Oxford ('B' is for *beorc*, 'birch', in the Old English runic alphabet). He wrote a poem in Gothic in praise of the birch in *Songs for the Philologists*, and another poem in that collection praises the birch and the

'B-scheme' together – the last 'B-scheme' graduates took their BAs in 1983. In that poem the derided opponent of the 'B-scheme' is the modern-literature-only 'A-scheme', represented by the oak (for in Old English 'A' is for *ác*); and Nokes's name is really Okes, a modern mispronunciation of the common place-name *æt þæm ácum*, later '*Atte(n) okes*'. But if one makes this admittedly obscure and personal connection, then the naked and weeping birch in Faery to which Smith offers to 'make amends', only to be told 'Go away and never return', becomes ominous. It is as if Tolkien still, in some way, felt guilty about stripping philology for his own purposes – in Niggle's terms, converting the potato-patch into a picture-shed.

Whatever one's detailed reading of the story, it is in general clear that *Smith of Wootton Major* is another 'Valedictory Address', or 'Farewell to Arms', in which Tolkien lays aside his star; defends the real-world utility of fantasy; insists that fantasy and faith are in harmony as visions of a higher world; hopes for a revival of both in a future in which the Nokeses of the world (the material-ists, the misologists) will have less power; and possibly, though this is my last and most tentative suggestion, expresses a veiled regret at his own denuding of the philological birch (a regret and a guilt which I share, see p. xi above). One cannot avoid noting, furthermore, that *Smith* picks up several motifs from his earlier shorter pieces, not all of them evidently meaningful. As Alex Lewis pointed out in 1991, the geography of *Smith of Wootton Major*, though reduced, is much the same as that of the Little Kingdom in *Farmer Giles of Ham*. Its few place-names may well be carefully chosen. 'Wootton' for instance means no more than *wudu-tun*, 'village in the wood', but one may see woods as portals of entry, with Verlyn Flieger, or as the very heart of 'tree-tangled Middle-earth' (see pp. 202–6 above). Smith also has an oddly consistent nomenclature (Nokes, Nell, Nan, Ned) marked by 'nunnation' – putting 'n' on the front of common words and names like Ann or Edward – just as the names in 'Leaf by Niggle' were marked by name-diminutives, Tompkins, Atkins, Perkins. I have suggested one possible implication of this for the name 'Nokes' just above. Major motifs, meanwhile, are 'the man who sees into Elfland', and 'the mortal returned to mortality', as in so many of Tolkien's poems, and his 'Lost Road' pieces. With so much of his personal life poured into it, one can see why Tolkien might strongly resist any sense that what he had written was a simple allegory, to be reduced to one all-inclusive meaning, a habit he thought much too common in schools of literary criticism. Nevertheless *Smith of Wootton Major* is quite clearly not just a surface narrative. Even more than usual for Tolkien, its extraordinarily simple style is deliberately deceptive.

As is proper for valedictory addresses, Tolkien was wrapping things up, looking back over his life (as over Smith's), taking the opportunity to make a final statement. One might compare *Smith*, finally, with 'Bilbo's Last Song', given by Tolkien to his secretary Joy Hill on 3rd September 1970, as a token of gratitude, but not published till 1974, the year after he died. Bilbo is also saying 'farewell' to his friends and to Middle-earth, but he is about to take the Lost Road, to make the Great Escape. His words could, however, entirely appropriately for myth, be removed from their 'Grey Havens' context and be heard as the words of a dying man: but one dying contented with his life and what he had achieved, and confident of the existence of a world and a fate beyond Middle-earth.

AFTERWORD

⋯≡◉≡⋯

THE FOLLOWERS AND THE CRITICS

Tolkien and the critics

In the preceding chapters I have tried, here and there, to mention
the serious criticisms of Tolkien which have been made sufficiently openly to enable a reply – criticisms of his morality,
style, characters, and narrative method (see, for instance, pp. 117,
147, 158, 224). A subject which has however for the most part
been shunned since the opening pages has been the general
phenomenon of intense critical hostility to Tolkien, the refusal
to allow him to be even a part of 'English literature', even on the
part of those self-professedly committed to 'widening the canon'.
One reason is that while the hostility is open enough, the reasons
for it often remain unexpressed, hints and sneers rather than
statements. Many critics are very ready to express their anger, to
call Tolkien childish and his readers retarded; they are less ready
to explain or defend their judgements. The assumption seems to
be that those of the right way of thinking (Susan Jeffreys's *literati*)
will know without being told, and those of the other party do
not deserve debate: classic tactics of attempted marginalization.
Recurrent features have been wild prediction (silly, because easily
proven false by later events), and self-contradiction (which is

self-revealing). Thus the anonymous reviewer of *The Fellowship of the Ring* for the *Times Literary Supplement* back in 1954 – we now know he was the historical novelist Alfred Duggan – predicted confidently, 'This is not a work which many adults will read through more than once'. It must have seemed a safe prediction at the time – few adults read any work as long as *The Lord of the Rings* even once, let alone more than once – but it was wrong, indeed it could not have been more wrong: of all popular bestsellers, *The Lord of the Rings* is the one most likely to be read over and over again. In the same way Philip Toynbee a few years later in 1961 (a friend of Duggan's, and another member of the literary coterie to be discussed below as the *Sonnenkinder*) made the equally confident prediction, quoted already on p. xx above, that the wave had passed, that Tolkien's supporters were beginning to 'sell out their shares', that the whole craze was passing into 'merciful oblivion'. It had not in fact got properly started, for popularity in the USA did not take off till the 'pirate' Ace edition of 1965 and the authorized Ballantine one that followed it in the same year.

In any case Toynbee in 1961 was contradicting himself with a curious mixture of insight and blindness. Just a few months before he had written a piece in the *Observer* for April 23rd, 'The Writer's Catechism', defining his image of 'the Good Writer'. The Good Writer he declared (N.B., the masculine pronouns in what follows are all Toynbee's) is a private and lonely creature who takes no heed of his public. He can write about anything and make it relevant, even 'incestuous dukes in Tierra del Fuego'. He 'creates an artifact which satisfies him' and 'can do no other'. When his work appears it will be 'shocking and amazing ... unexpected by the public mind. It is for the public to adjust'. Nearly all of this is a perfect description of Tolkien: 'private and lonely', writing in his converted garage; concerned only with his artifact, or one might say his Tree; his work utterly unexpected on publication, and yet capable of making anything 'relevant', even fantastic creatures in an invented world. And when one adds to Toynbee's list his *coup de grâce*, his crowning identifier, that 'the Good Writer is not directly concerned with communication, *but with a personal struggle against the intractable medium of modern English*' (my emphasis), it is hard to see how he could miss the connection: Tolkien saw deeper into and reacted harder against the nature of specifically modern English than any other writer this century, as can be said without qualification.

And yet Toynbee was not alone in this strange inability to see what (he said) he was looking for. In 1956 Edmund Wilson, then doyen of American modernist critics, had dismissed *The Lord of the Rings* as 'balderdash', 'juvenile trash', a taste which he thought was specifically British (once again a prophecy about to crash in flames as the American market conversely took off). In his 1931 critical classic *Axel's Castle*, however, he had sternly if pompously rebuked exactly this tendency towards dismissal:

> it is well to remember the mysteriousness of the states with which we respond to the stimulus of works of literature and the primarily suggestive character of the language in which these works are written, on any occasion when we may be tempted to characterize as 'nonsense', 'balderdash' or 'gibberish' some new and outlandish-looking piece of writing to which we do not happen to respond. If other persons say they do respond, and derive from doing so pleasure or profit, we must take them at their word.

The last sentence could not be put better. But when the event happened, Wilson was first in line with 'balderdash', exactly the word he had outlawed. He had completely forgotten his own rule.

What is the psychology of this, one wonders? Do these people not mean what they say? And why can't they say what they mean? Another feature of response to Tolkien has been what I can only call simple snootiness, and what Orwell called the 'automatic snigger' of the English-speaking Establishment intellectual. Susan Jeffreys's remark about '*literati*' was mentioned at the start of this book. It is matched by Anthony Burgess twenty years before (in the *Observer* for 26th November 1978) dismissing 'allegories with animals or fairies' in favour of the 'higher literary aspirations'. By the allegories he meant, I think, *Watership Down* and *The Lord of the Rings* (neither of them an allegory) – I doubt he would have had the nerve to castigate *Animal Farm*, which certainly is an allegory with animals. But what he meant by the 'higher literary aspirations' he does not say: if we were *literati*, presumably, we would know already. I am never able to refrain from citing Professor Mark Roberts's judicious and total rejection of *The Lord of the Rings* – as suggested above, p. 156, possibly the single most imperceptive statement ever made about Tolkien:

> It doesn't issue from an understanding of reality which is not to be denied, it is not moulded by some controlling vision of things which is at the same time its *raison d'être*.

If there is one work of which one can be sure this *is* true, it is *The Lord of the Rings*, indeed it would be possible to criticize it for its author's utter single-mindedness. But somehow Roberts, like Toynbee, like Wilson, missed seeing it. They were looking for a literary revelation, but when one came they denied it. It was not what they expected. Populist, not élitist. It did not provide that comfortable sense of superiority to the masses without which the English-speaking literary intellectual, it seems, cannot cope at all (a point made in much more detail by John Carey's iconoclastic book from 1993, *The Intellectuals and the Masses*).

Several attempts have been made to explain this deep and seemingly compulsive antipathy. I have suggested in *The Road to Middle-earth* that it is at bottom professional, a reflex of the ongoing language/literature war which has preoccupied university departments of English for a century: this is perhaps too narrow an explanation. Patrick Curry argues in his *Defending Middle-earth* that it stems from a kind of generation war, as a group devoted to 'modernism', and to thinking themselves up-to-date, find themselves pushed aside by 'postmodernism': the argument gains a lot of force from Tolkien's surging popularity with protest movements in the West and even more in Eastern Europe, but Curry's definition of 'postmodernism' is a personal and tactical one (he considers the 'hostility' phenomenon much more widely in his 1999 article). Joseph Pearce's 1997 book *Tolkien: Man and Myth* implies that the antipathy is a reaction if not to Tolkien's Catholicism specifically, then to his 'religious sensibility': again not impossible, but not often overtly mentioned. That there is a class basis to the critics' reaction is strongly suggested, for instance, by Humphrey Carpenter's contemptuous dismissal of Tolkien's followers as 'anorak-clad' (cited by Pearce): obviously Carpentercannot know what proportion of Tolkien-readers habitually wear anoraks, nor would it say anything about their literary tastes if he did. But the reference to anoraks is easily understood, is intended to be understood, as class-hostility from those who habitually carry umbrellas: a very clear case of the *haute bourgeoisie* insisting on retaining its monopoly of culture. To balance matters, Jessica Yates points out that Tolkien often triggered extreme hostility from consciously left-wing academics (though the *haute bourgeoisie* is often theoretically left-wing, see the account of the *Sonnenkinder* below).

There is some truth in all the theories just proposed, nor do they necessarily exclude each other. However, they are all in their way arguments from off the page. It would be better if some kind of literary case could be made out to explain this curious phenomenon of seemingly irrational hatred.

Tolkien and Joyce

Comparing Tolkien and Joyce, *The Lord of the Rings* and *Ulysses*, may well inflame the situation rather than cool it. To critics like Germaine Greer, whose view is cited on p. xxii above, the comparison would probably be all but blasphemous. However, the general opinion that Tolkien knew nothing of literary history, was unswervingly hostile to Shakespeare and Milton and the entire post-medieval canon, has been shown to be false in chapter IV. I can see no sign that he read or admired Joyce. Nevertheless, something can be learned by putting the two men, and the two works, into relation with each other. They were, after all, both authors of the same century, close contemporaries, not dissimilar in background.

There are some immediate points of similarity. The two men's careers are more like each other's than most other major writers: each with one obvious main work, each of these to some extent a development of an earlier and shorter one with which it shares some characters (*The Hobbit*, *Portrait of the Artist*), the two sequences supported only by some short pieces and by collections of poems, and extended by posthumous publication of first drafts (like Joyce's *Stephen Hero*, and the volumes of the *James Joyce Archive*). It is true that Joyce's *magnum opus* came out when he was only forty, while Tolkien waited till he was sixty-two. Tolkien might have replied that he did not have the massive financial support Joyce received for his writing – it has been calculated, some £23,000 between 1915 and 1930, certainly more than Tolkien earned from being a professor in the same period. Joyce was in fact in the situation that Niggle could only dream about: in receipt of a pension given to him entirely so that he could get on with his writing, with no garden to neglect and no threat of an Inspector.

But there are less accidental connections. Joyce never achieved

Tolkien's academic stature nor his learning, but he was a philologist of sorts. We know he took a course in the subject at University College, Dublin; the 'Oxen of the Sun' sequence in *Ulysses* shows him putting it to use; and the 'Proteus' section is assigned overtly to 'Philology' in Joyce's own scheme for the book. Probably Joyce was the kind of 'Good Writer' Philip Toynbee was thinking of when he wrote about personal struggles against 'the intractable medium of modern English'. More subtly, both *Ulysses* and *The Lord of the Rings* are evidently works of the twentieth century, neither of them readily describable as novels, which are engaged in deep negotiation with the ancient genres of epic and romance (the structure of *Ulysses* parallels that of the *Odyssey*, generally agreed to be the more romantic of the two epics ascribed to Homer). More comically, both of them got something of the same treatment from sections of the intelligentsia when they eventually appeared. The class-reactions to Tolkien are noted above, but one might compare with them Virginia Woolf's nettled dismissal of Joyce's work in her diary as 'illiterate, underbred'. 'Underbred' is exactly the same sort of sneer as 'anorak-clad' – the *haute bourgeoisie* insisting on its monopoly of culture again – but what *could* she have meant by 'illiterate'? No doubt, 'not pleasing to the *literati*', though as far as Joyce was concerned she was wrong even about that. The joking dismissal of the *Silmarillion* as 'a telephone directory in Elvish' has been mentioned above (p. 242). It may add to the joke that *Ulysses* has been shown to depend heavily on just that, the 1904 edition of a Thoms's Dublin *Directory*. Both works are deliberately schematized, as we know because their authors' schemes survive; and they are also very clearly in intention encyclopaedic.

The differences, of course, are even more striking than the similarities. The action of *Ulysses* is confined within a single day, 16th June 1904, and a single city, Dublin: the scope of *The Lord of the Rings* is both historically and geographically much greater. Indeed it is not too much to say, and it is not even a criticism to say it, that in the main action of *Ulysses* nothing much happens. It could be called, 'One Day in the Life of a Nobody'; and such a title makes possible a further comparison. Solzhenitsyn famously wrote another work about a nobody, very nearly a non-person, *One Day in the Life of Ivan Denisovitch* (1963); but it is nothing like *Ulysses*. Solzhenitsyn takes one day in one life as a sample, significant because it is just like so many millions of other days in other lives. The whole point of the work is a public one, bitterly satirical, aggressively political. *Ulysses* by contrast is nothing if not private and personal. What it shows above all is the complexity and individuality even in the inner life of a nobody. It becomes at times literally a Babel of voices, but many of those voices are the same voice, from a self which is intrinsically heterogeneous. To it (T.S. Eliot suggested) the only possible response is silence – perhaps because another of its distinctive features is that, unlike *The Lord of the Rings*, it refuses to follow even the most conventional of plot outlines. E.M. Forster (not much of an artist with plots himself) observed that in most novels there is a moment when complication turns towards resolution, and in *The Lord of the Rings* one might well be able to pinpoint exactly when this is – when Ghân-buri-Ghân cries out 'Wind is changing!', perhaps (V/5) or maybe, closer to the physical centre, when Gandalf says 'The great storm is coming, but the tide has turned' (III/5). There is no sign of any such centre-point or change of direction in *Ulysses*. Its flux continues to the end.

Tolkien and modernism

Rather the same phenomenon of superficial similarity and deeper opposition appears if one widens the scope of the argument to considering the whole phenomenon of 'modernism', of which *Ulysses* is accepted as a definitive work. Authoritative recent accounts of modernism – I use here primarily the entries in *The*

Oxford Companion to English Literature compiled by Margaret Drabble (1998), and in Michael Groden and Martin Kreiswirth's *Johns Hopkins Guide to Literary Theory and Criticism* (1994) – often seem immediately applicable to Tolkien. Modernist style, we are told, is characteristically local, limited, finding beauty not in abstractions but in 'small, dry things'. I am not sure about 'dry', but Tolkien appears to present himself in 'Leaf by Niggle' as essentially a miniaturist (see p. 268 above), and the impression is confirmed not only by the many passages of close natural description in all his works (the 'purple emperor' butterflies in Mirkwood, the falling willow-leaves by the Withywindle), but by, for instance, his long, careful, deeply absorbed study of hybrid flowers from a single plant in his garden (see letter to Amy Ronald in *Letters*). Modernism was said furthermore, by T.S. Eliot, to have made it possible to replace narrative method by 'mythical method'; and the whole drive of Tolkien's work, as one can see, was towards creating a mythology which his major narrative was there to embody. When one reads also (this time in Drabble) that modernism is distinguished by experiments with the representation of time; by rejection of the 'realist illusion'; by the use of multiple narrators; and by experiments with language, one might well check them off by remarking, respectively: 'yes, see 'The Lost Road' and 'The Notion Club Papers'; the experiments with interlaced narrative, the use of 'threads' of story alternating and contrasting; and, of course, the deliberate creation of unknown languages and unrecorded dialects'. As for a liking for irony, also cited as characteristically modernist, Tolkien's whole developed narrative method is ironic, as also anti-ironic (see pp. 110–11 above). Why is it unacceptable to see Tolkien, then (12 lines in the Drabble *Companion*), as a modernist author parallel to Joyce (76 lines), and a 'Good Writer' of exactly the sort imagined by Toynbee?

The answer is clear enough if one looks at some of the other features itemized. Modernist works tend to rely very heavily on literary allusion – as, for instance, in Eliot's 'Waste Land' or Joyce's *Ulysses*. If the reader does not follow the allusion, does not realize the contrast between the words in their original context (in Homer, say, or Dante) and in their modernist context, then the point is lost. Tolkien by contrast was as well read as anyone and more so than most, and he alludes frequently to works of what he regarded as his own tradition, the 'Shire tradition' of native English poetry. It is absolutely characteristic of his uses of tradition, however, that the source of the allusions *does not matter*. The words work best when they have become quasi-proverbial, common property, merged with ordinary language, 'as old as the hills'. Many of the works he used most are anonymous. Tolkien never subscribed to the cult of the Great Author, the person raised above the common clay, so evident for instance in E.M. Forster's again definitive short story 'The Celestial Omnibus' (a work deliberately parodied, I believe, in Lewis's study of death already mentioned, *The Great Divorce*). Though he started by reading Classics at Oxford, Tolkien was also determinedly hostile to 'the Classical Tradition', as Eliot called it. Joyce's schema depends on Homer, Eliot alludes continually to the tales of Agamemnon and Tiresias, Oedipus and Antigone. Milton attempted to supersede these (though he knew them better than anyone alive) by the heroes of the Bible. But Tolkien's heroes and his major debts came from the native and Northern tradition which Milton never knew and Eliot ignored: Beowulf, Sir Gawain, Sigurð, the Eddic gods – a tradition seen by most modernists as literally barbarous (the possession of people who speak incomprehensible languages).

A final contrast is the modernist love of introspection, of the 'stream of consciousness' technique, of the characteristic trick of even the simplest of modern novels of telling you *what the characters are thinking*. Tolkien does this too, in *The Hobbit* and *The Lord of the Rings* (much more rarely in *The Silmarillion*); it may be impossible to present a narrative successfully to modern readers without it, though this is an experiment which no devotee of 'experimental fiction' has to my knowledge tried. Tolkien was however well aware of works which had tried it, and tried it successfully. There is only one moment in *Beowulf* when the narrator hesitates, as it were, on the brink of entering his character's mind, when he says of his hero, 'his breast boiled with dark thoughts, as was not his custom', because the dragon has just burned down his hall and he does not know what he has done to deserve it. But the poem goes on, 'He ordered a splendid war-shield to be made all of iron ... he knew wood would not help him, linden-wood against fire', with no more words wasted on the dark thoughts. Sir Gawain has dark thoughts too, but they are allowed out only when he mutters in his sleep (Tolkien's translation), 'as a man whose mind was bemused with many mournful thoughts'. We never learn what they are, because Gawain too resumes his public face immediately. In the cultures Tolkien admired, introspection was not admired. He was aware of it, in a way his ancient models were not, but he did not develop it.

Once one sees this utter opposition of literary philosophy, even the superficial similarities listed above are exposed. Tolkien's approach to the ideas or the devices accepted as modernist is radically different because they are on principle *not literary*. He used 'mythical method' not because it was an interesting method but because he believed that the myths were true. He showed his characters wandering in the wilderness and entirely mistaken in their guesses not because he wanted to shatter the 'realist illusion' of fiction, but because he thought all our views of reality *were* illusions, and that everyone is in a way wandering in a 'bewilderment', lost in the star-occluding forest of Middle-earth. He experimented with language not to see what interesting effects could be produced but because he thought all forms of human language were already an experiment. One might almost say that he took the ideals of modernism seriously instead of playing around with

315

them. But what he forfeited in the process – like Bilbo Baggins, who was never regarded as respectable again – was the underlying and, one has to say, always potentially snobbish and élitist claim of so much modernist writing, that it was produced for and could only be appreciated by the thoroughly cultivated individual, the fine and superior sensibility.

This is probably at the heart of the critical rage, and fear, which Tolkien immediately and ever after provoked. He threatened the authority of the arbiters of taste, the critics, the educationalists, the *literati*. He was as educated as they were, but in a different school. He would not sign the unwritten Articles of the Church of Literary English. His work was from the start appreciated by a mass market, unlike *Ulysses*, first printed in a limited number of copies designedly to be sold to the wealthy and cultivated alone. But it showed an improper ambition, as if it had ideas above the proper station of popular trash. It was the combination that could not be forgiven.

The lack of perception shown by Philip Toynbee and Alfred Duggan is, finally, interesting in more than one way, for both were members of the literary coterie which ruled and defined English literature at least for a time, between the wars and after World War II, and which the critic Martin Green has called, in his 1977 book *The Children of the Sun*, 'the *Sonnenkinder*'. They were committed modernists, upper class, often Etonians, often professed Communists, often (like Duggan) extremely rich, well-entrenched as editors and reviewers in the literary columns, though with a pervasive problem of failure to produce – Toynbee's mentor Cyril Connolly's one classic work remains his long excuse for failure, *Enemies of Promise*. Their leading literary figure was Evelyn Waugh (also high up the Waterstone's rankings), whose son Auberon continues to figure prominently in the attacks on Tolkien. In the 1960s, when Toynbee was writing, though they still in Green's phrase 'staffed the Establishment', they were becoming *passé*, a dreadful if ultimately inevitable fate for those committed to being *avant-garde*: this accounts for the venom of some of their attacks.

However, I would give at least the next-to-last word on this subject to Martin Green, a conspicuously fair-minded writer with almost no interest in Tolkien (whose name in early editions of his book he invariably misspells). The Inklings, he wrote – Charles Williams and Dorothy Sayers, Lewis and (sic) Tolkein – avoided the poses of the *Sonnenkinder*, and centred their thinking on orthodox Christian theology, and on the problem of evil. Green admits that 'Most aspects of their ideological and imaginative behaviour' strike him as:

> more generous, intelligent and dignified than those of either Leavis [doyen of mid-century English critics] or Waugh – or Orwell, for that matter, if considered in the abstract. But considered in the concrete, the ideas of the last three have at various times meant everything to me, while the others *mean*, in that sense, nothing. I approve what they did, but theoretically; I read the books it resulted in approvingly, but I am not really at all engaged by them.
>
> And one reason surely is that these writers removed themselves from the cultural dialectic. Undignified as that often was, both personally and intellectually, that was where the action was . . .
>
> (Green, 1977, pp. 495–6)

I understand and respect Green's position, though it is not mine. His last remark however reminds me of a famous music-hall joke, a kind of sub-literary *Waiting for Godot*. On a darkened stage, a single light is burning. A man is down on his hands and knees, crawling round in silence, obviously looking for something. Eventually a second man comes on, and says, after watching for a while, 'What are you doing?' 'I'm looking for a sixpence I dropped', replies the crawler. The second man gets down on his hands and knees and starts to help him. After a while the second man says, 'Just where did you drop it, anyway?' 'Oh, over there', says the first one, getting to his feet and walking over to the other side of the stage, in the dark. 'Then why are you looking for it here?' cries the second man in exasperation. The first one walks back to his original place and starts crawling round again. 'Because', he replies, 'that's where the light is'.

In this allegory of mine, the light = modernism, the crawling searcher = Toynbee (or Greer, or Susan Jeffreys from the *Sunday Times*, any of the critical multitude). I am not at all sure what the sixpence may =, but Tolkien was out there in the dark, looking for it.

Tolkien's legacy

It is a relief to turn from hate and fear to love and admiration. Any full study of Tolkien's many imitators would have to be at least book-length – incidentally, and to be fair, the entry on 'Fantasy Fiction' in the Drabble *Companion* is just as long as the one on 'Modernism': modesty prevents me from recommending it. But there is some interest in recording what a few of his most evident emulators have found inspirational in Tolkien, as in noting what they leave alone, or cannot approach.

At the most elementary level, reading Tolkien produced a strong desire for more stories about hobbits – a desire which Sir Stanley Unwin had identified as far back as 1937. Writing stories about hobbits pure and simple has remained difficult, however, as hobbits (in spite of *The Denham Tracts* and the *OED*) remain so clearly a Tolkien invention. Various evasions have been tried: there is an anthology called, rather unfortunately, *Hobbits, Halflings, Warrows, and Wee Folk*; the Martin Greenberg anthology of stories in honour of Tolkien, *After the King*, contains Dennis McKiernan's 'The Halfling House'. None of these efforts is very successful at catching the hobbit flavour, of course increasingly anachronistic even in its 'modern' or Edwardian aspect, especially for American writers and readers.

At a slightly higher level, some fans seem just to want to write (and read) *The Lord of the Rings* all over again. In Diana Wynne Jones's excellent and not at all Tolkienian fantasy *Fire and Hemlock* (1984), the girl-heroine discovers *The Lord of the Rings* at the age, seemingly, of about fourteen, and reads it through four times running. She then immediately writes an adventure story about herself and her own mentor/father-figure:

> how they hunted the Obah Cypt in the Caves of Doom, with the help of Tan Thare, Tan Hanivar, and Tan Audel [the other members of her mentor's string-quartet]. After *The Lord of the Rings* it was very clear to her that the Obah Cypt was really a ring which was very dangerous and had to be destroyed. Hero did this, with great courage.

But when she sends her story to Tam Lynn, he only writes back, '*No, it's not a ring. You stole that from Tolkien, use your own ideas*'. The deflating comment seems appropriate to a good deal of Tolkien imitation, whose drive is to have the same thing again, only more of it.

The most obvious example is Terry Brooks's generally derided, but still commercially successful, *The Sword of Shannara*. Rumour has it that when this came out in 1977 it had been commissioned by astute editors who knew they could sell anything sufficiently Tolkienian. If so, the editors were right. The 'Shannara' sequence is still running twenty years later, and is up to eight volumes. Yet the strange thing about the first volume at least is the dogged way in which it follows Tolkien point for point. A group is assembled to retrieve a talisman from the power of a Dark Lord. It is 'retrieve', not 'destroy', which is one point of dissimilarity. But the group assembled matches Tolkien's Fellowship very nearly person for person. There is a Druid, or wizard, Allanon (= Gandalf); a dwarf, Hendel (= Gimli); two youths, central characters, who take the place of the four hobbits; two elves, one more than Tolkien's Legolas, but then one of them is called Durin, a Tolkien name; and two men, Menion and Balinor, corresponding closely (Balinor too has a younger brother) to Aragorn and Boromir. Gollum is reincarnated in the person of Orl Fane, a gnome who gets possession for a time of the Sword of Shannara and dies trying to regain it. The Ringwraiths re-appear, 'deathlike cry' and all, as flying Skull Bearers, while the phial of Galadriel is replaced as a weapon against them by the Elfstones. As if that were not enough, the plot-outline is followed very nearly point for point as well: first journey to a 'homely house', Culhaven = Rivendell; pause in a hallowed forest, Starlock = Lórien; loss of Allanon, who is dragged into a fiery pit by a Skull Bearer, just like the Bridge of Khazad-Dûm (though like Gandalf he re-appears); and even, ambitiously though on a very small scale, the separation of the company when the hobbit-analogues are captured and led away by orc-analogues, only to be re-united later (after the expected tracking-scene). There are analogues to Sauron, Denethor, Wormtongue. The hobbit-analogues are attacked by 'Mist Wraiths' (like the barrow-wight), a tentacled creature in a pool (like the Watcher by Moria-gate), by a malevolent tree (Willow-man). Individual scenes are quite closely imitated, like the slamming down of the stone door at the end of *The Two Towers*, the death and withering of Saruman, or the arrival of the Riders of Rohan on the Pelennor Fields. The similarity is so close that in a way it is hard to tell how good or bad the result is. Anyone who had not read *The Lord of the Rings* might find it highly innovative – but I doubt that many of its original readers fell into that category. What *The Sword of Shannara* seems to show is that many readers had developed the taste (the addiction) for heroic fantasy so strongly that if they could not get the real thing they would take any substitute, no matter how diluted.

This is not the case with Stephen Donaldson's 'Thomas Covenant' series, a work generally agreed to be much more original, and to have become in the end something like a critique and even an attempted rebuttal of Tolkien (see the entry for Donaldson in Clute and Grant's *Encyclopedia of Fantasy* already mentioned, and the full-length study of Donaldson by W.A. Senior). Nevertheless, the Tolkien impression is there, deep-stamped. A major and deliberate difference is that this time the central character is nothing like a hobbit, is in fact a modern adult American, who happens however to be a leper and becomes a rapist (about as far from Bilbo and Frodo as one could get). Nor, this time, does the anti-hero accumulate a Fellowship, as in the Brooks imitation. The similarity between Tolkien and Donaldson is rather in the landscape, or the people-scape, through which the anti-hero passes. The first volume in the sequence, *Lord Foul's Bane* (also 1977), starts with a Cavewight recovering a talisman (like Gollum with the Ring), with in the background a story of a maimed hero, Berek Halfhand (cp. Beren the One-Handed). The Ringwraiths re-appear as 'Ravers' (not a fortunate choice of name); invocations of Elbereth are paralleled by invocations of Melenkurion ('You have spoken a name which no Raver would call upon', Donaldson, cp. Sam Gamgee, '*Elbereth* I'll call. What the Elves say. No orc would say that'); the tree-houses of Lórien figure once again as 'Soaring Woodhelven', the wood of the strangling Huorns comes back as the forest of 'Garroting Deep', a troop of riders appears as 'the Third Eoman' (Éomer's was the third *éored*), there is even a biting-off-finger scene. Especially close is the invention of the Giant Saltheart Foamfollower and his people, who correspond to Tolkien's Ents once again virtually point for point: Saltheart looks 'like an oak come to life', has 'deep-set eyes' which 'flashed piercingly, like gleams from his cavernous thoughts'; and he sings 'in a language Covenant could not understand', explains that Giantish is hard to translate, because the Giants' tales take too long to tell, and regrets that 'we have so few children'.

Yet Donaldson has said, and I for one believe him:

> Tolkien influenced me powerfully by inspiring in me a·
> desire to write fantasy. But when I actually began writing
> the Covenant books, I stayed as far away from Tolkien's
> example as the exigencies of my own story allowed.
>
> (Senior, *Donaldson*, p. 250)

Reconciliation of the observed facts and the author's statement
may be gained by noting that Donaldson uses several words which
were to say the least extremely uncommon (especially in America)
before being used by Tolkien: for instance, 'gangrel', 'eyot', and
'dour-handed', the latter surely a borrowing. Yet people often do
not remember where or when they learned particular words, nor
do they regard them as a debt. My suggestion is that in some
cases – many cases, like Diana Wynne Jones's heroine – Tolkienian words, and images, are learned so early and so thoroughly,
possibly by compulsive re-reading, that they become internalized,
personal property rather than literary debt. The phenomenon
was common enough in the days of ballad culture or oral-
formulaic epic; passive bearers of a tradition merged readily with
active extenders of it. It is a strange but not entirely unwelcome
thing to see in an age of individual authorship and defended
copyright.

My last example of the relationship between Tolkien and later
admirers is again a first novel, *The Weirdstone of Brisingamen* by
Alan Garner (1960). Garner is at once the most like and the most
unlike Tolkien of the authors mentioned here. Garner is English,
has written several 'young adult' novels of great distinction and
originality, and a recent adult novel, *Strandloper*. He is a native of
Cheshire, and most of his books centre on Alderley Edge, as personal and as full of mythic potential to him as the West Midlands
were to Tolkien. In *Strandloper* Garner interweaves overt and
overt quotations from the works of the *Gawain*-poet, whom he

and most critics (myself not included, see p. 199 above) take to be also a Cheshire man. *The Weirdstone of Brisingamen*, then, is set on Alderley Edge, in modern times, though it starts off from an old local legend. It has no hobbits, its central characters being two children. But like Frodo they have come by accident into the possession of a vital talisman being sought for by a Dark Lord who (a reversal of the Tolkien theme) wishes to destroy it to put an end to the protective magic of the white wizard, Cadellin. In their association with Cadellin the children find themselves in contact with dwarves, a troll, and orc-analogues called 'the morthbrood' ('orc' is another word very closely identified with Tolkien, though not quite as definite an invention as 'hobbit'). The similarities of plot are not strong (unlike Brooks), the book's personnel could come as easily from traditional fairy-tale as from Tolkien's re-creation of it (unlike Donaldson). Yet the Tolkienian influence remains pervasive, on the level of scene and even more of phrase – something which could well again be unconscious. Fenodyree the dwarf tells the children, as they crawl through the tunnels:

> 'so deep did men delve that they touched upon the secret places of the earth ... There were the first mines of our people dug, before Fundindelve: little remains now, save the upper paths, and they are places of dread, even for dwarfs.'

It seems a half-memory of Gandalf speaking of Moria, the Dwarrowdelf, 'too deep they delved, and woke the hidden dread'. In similar fashion Cadellin says of an earlier defeat of *The Weirdstone*'s 'Dark Lord' that when he fled 'all men rejoiced, thinking that evil had vanished from the world for ever'; but he has returned, 'pouring black thoughts from his lair in Ragnarok'. Elrond also remembers the time 'when Thangorodrim was broken, and the Elves deemed that evil was ended for ever, and it was not so', and Gandalf corroborates him, saying that already in Bilbo's time the Necromancer was 'sending out his dark thought from Mirkwood'. The 'thin cry, like the plaintive voice of a night-bird, yet cold and pitiless as the fangs of mountains' is in Garner that of a she-troll, but it is like the 'long-drawn wail ... the cry of some evil and lonely creature' which signals a Ringwraith. The children are tracked by black crows in scenes strongly reminiscent of the spying *crebain* in 'The Ring Goes South'.

What Garner has learned here is perhaps only the trick of varying style, allowing a proportion of archaic (or in Garner dialectal) language to tinge some characters' speech and some narration, to make it strange but still comprehensible; and there are some things he does, reaching past Tolkien to Tolkien's own sources, so to speak, which are quite un-Tolkienian. I do not think Tolkien would have approved, for one thing, of Garner's trick of taking genuine Old Norse words out of their context and using them just as names, like Nastrond – in Garner, the name of the Dark Lord, but in Snorri's *Edda Náströnd*, the 'Corpse-beaches', the place where sinners go after death – or Ragnarok – in Garner, again, the Dark Lord's stronghold, but in Norse *Ragna-rök*, Doomsday, the Destruction of the Gods. *Brisinga men* (two words) is the name of the goddess Freya's necklace in Norse: Garner uses it only for strangeness. And yet in making these borrowings Garner is following a Tolkienian theory, that people can tell the genuine from the fake, even when it comes to making up names. Do not make them up, therefore. If you cannot embed them in a language (like Quenya or Sindarin), borrow them from an existing language. Garner shows a certain respect for his predecessor even in disagreement or deviation.

Having looked at what authors *have* taken from Tolkien, consciously or unconsciously, it may be worth finally considering what they have *not*. Has Tolkien proved to be in any way like his popular predecessor Dickens, 'the Inimitable'? One interesting feature which no one has attempted to copy in any detail has been Tolkien's continual insertion of poems, in very different styles and often complex metres. It could be that this is too much trouble, but another factor is probably the sheer depth of Tolkien's involvement with literary tradition: fantasy writers are not brought up the way he was any more. Along with this goes a lack of interest in literary gaps, errors, contradictions. Fantasy authors are very ready to raid works like the *Elder Edda* or *Sir Gawain and the Green Knight* for material, but not to rewrite them, point out their mistakes, 'reconstruct' narrative which is no longer there. The collapse of philology as a university discipline makes it likely that this will remain permanent. A further feature which as far as I know no one has ever tried seriously to copy is Tolkien's structuring of *The Lord of the Rings*, his use of narrative threads. For one thing the very careful chronological positioning, the cross-checking of dates and distances and phases of the moon would be hard to do accurately, something best left to a 'natural niggler'. For another, it seems likely that no modern author feels able to accept Tolkien's highly Boethian ideas of fortune, chance and Providence, even when checked and balanced by the anti-Boethian suggestions also present. The underlying sadness of his work, its many death-scenes and avoidance of the unmodified happy ending, presents a further challenge to the world of commercial publishing.

Nevertheless, it would not be true to say that Tolkien's imitators have responded only to surface features. Once again, as far as I know no contemporary writer has gone as far as Tolkien did in the creation of imaginary languages, and probably no one could, but what have been very well digested are his views about the importance of language, the importance of names, and the necessity for a feeling of historical depth. I think Tolkien himself would have had to raise an appreciative eyebrow at the amount of linguistic knowledge packed into the works of, for instance, Avram Davidson. I remarked in the Foreword that not everyone responds to Gothic, but Davidson, in his *Peregrine Secundus*

(1982) expects his readers to relish Ancient Oscan, and clearly many of them do. Michael Swanwick's brilliantly inventive *The Iron Dragon's Daughter* (1993) meanwhile seems to me to show that the author has been reading, if not *The Denham Tracts*, then something very like them. Authors like Swanwick, and Davidson, and Jack Vance, and many others including those mentioned below, value authenticity and what Tolkien called the 'flavour' that '*rooted* works have', because they have been shown their value. No one, perhaps, is ever again going to emulate Tolkien in sheer quantity of effort, in building up the maps and the languages and the histories and the mythologies of one invented world, as no one is ever again going to have his philological resources to draw on. Just the same, modern authors of fantasy probably accept that they have to work much harder than their predecessors from the nineteenth century, the William Morrises and Lord Dunsanys.

Such parallels could be drawn out at great length, and could be applied, in different ways and to different extents, to writers such as George R.R. Martin, Michael Scott Rohan (his name is a coincidence), Robert Jordan, David Eddings, Guy Gavriel Kay (Christopher Tolkien's assistant with *The Silmarillion*), and literally scores of others. I do not think any modern writer of epic fantasy has managed to escape the mark of Tolkien, no matter how hard many of them have tried. Most of them would probably not see it as a mark, or would accept the word only in the sense of something to aim at. Naturally all of them want to write individually, and very often they do – the differences in basic philosophy between Tolkien and Donaldson have for instance been powerfully brought out by William Senior's study already mentioned. But one might still think that, like Diana Wynne Jones's girl-heroine at a much simpler level, what all of them want to do is to produce the same result, satisfy the same appetite, as that achieved or satisfied by Tolkien.

Relevance and realism

A final attempt may be made to put together the negative and the positive views of Tolkien glanced at in this Afterword. The bedrock reason for ignoring Tolkien and disliking fantasy may be the sense that it is just plain *not true*. An affecting statement of this position comes from the great realist George Eliot, in *Adam Bede* (1859). 'I am content to tell my simple story', she says, 'dreading nothing . . . but falsity'. Falsity is easier than truth, if more ambitious. And so:

> I turn, without shrinking, from cloud-borne angels, from prophets, sibyls, and heroic warriors, to an old woman bending over her flower-pot, or eating her solitary dinner.

Or, one might say, from Valar, Maiar, elf-lords and Rangers, to days in the lives of nobodies. Eliot's view is a strong and dignified one, but one could respond to it in three ways. First, of course, hobbits are at least as close to the old woman and her flower-pot as they are to prophets, sibyls, and heroic warriors. Second, Eliot makes the claim for truth and for a simple story, but she then goes on to write fiction. The argument that fantasy is intrinsically less truthful than realistic fiction could be extended to say that realistic fiction is intrinsically less truthful than biography. But we all (now) know that fiction allows a writer to express something, perhaps metaphorically or by analogy, which could not be expressed by history. The same argument should be extended to fantasy. That is surely why so many writers of the twentieth century, including the ones most closely concerned with real-world events, have had to write in the fantastic mode.

The final argument follows on from the above. One of the things that fiction allows its creators to do is to express pattern. One might say, 'create pattern', but it is clear that in many cases the authors believe that they did not create it, they merely perceived it, and are now trying to make it clear to others. This is true of George Eliot, whose *Silas Marner* (1861) is a work which follows exactly the same kind of Providential interlacing as we see in *The Lord of the Rings*, if on a much smaller scale, and which furthermore climaxes in a speech (the old country-woman Dolly Varden's) which is a deliberate dialect paraphrase of Boethius, dissimilar to Gandalf's statements only in style, not content. If this authorial patterning is acceptable and desirable in realistic fiction, why should the same freedom not be extended to fantasy? Both forms are literally 'not true, just made-up'. But no one has to read everything literally.

I believe that it is our ability to read metaphorically which has made Tolkien's stories directly relevant to the twentieth century. We do not expect to meet Ringwraiths, but 'wraithing' is a genuine danger; we do not expect to meet dragons, but the 'dragon-sickness' is perfectly common; there is no Fangorn, but Sarumans are everywhere. It may indeed be the readiness with which these points are taken which has made Tolkien seem, not irrelevant, but downright threatening, to members of the cultural Establishment. Be that as it may, what Tolkien certainly did was introduce a new, or possibly re-introduce an old and forgotten taste into the literary world. A taste, a trace-element, perhaps a necessary literary vitamin? Whatever one calls it, to use the words of Holofernes, Shakespeare's pedant-poet in *Love's Labour's Lost*, if not in the way that Holofernes meant them:

'The gift is good in those in whom it is acute, and I am thankful for it.'

LIST OF REFERENCES

WORKS BY J.R.R. TOLKIEN

The Hobbit, first edition London 1937, cited here from fourth
   edition, London: George Allen & Unwin, 1978; Boston:
   Houghton Mifflin, 1978
'Leaf by Niggle', first published in *The Dublin Review*, January
   1945, 46–61, cited here from *Tree and Leaf, Smith of
   Wootton Major, The Homecoming of Beorhtnoth Beorhthelm's
   Son*, London: Unwin Paperbacks, 1975
Farmer Giles of Ham, first edition London: George Allen &
   Unwin, 1949; Boston: Houghton Mifflin, 1950, cited here
   from *Farmer Giles of Ham, The Adventures of Tom
   Bombadil*, London: Unwin Paperbacks, 1975
'The Homecoming of Beorhtnoth Beorhthelm's Son', first
   published in *Essays and Studies* 6 (1953), 1–18, cited here
   from *Tree and Leaf* (etc.), above
The Lord of the Rings, in three volumes:
I, *The Fellowship of the Ring*, first edition London: George
   Allen & Unwin 1954; Boston: Houghton Mifflin, 1954, cited
   here from second edition, London: George Allen & Unwin
   1966; Boston: Houghton Mifflin, 1967
II, *The Two Towers*, first edition London: George Allen &
   Unwin 1954; Boston: Houghton Mifflin, 1955, cited here
   from second edition, London: George Allen & Unwin 1966;
   Boston: Houghton Mifflin, 1967
III, *The Return of the King*, first edition London: George Allen
   & Unwin 1955; Boston: Houghton Mifflin, 1956, cited here

from second edition, London: George Allen & Unwin 1966; Boston: Houghton Mifflin, 1967

The Adventures of Tom Bombadil, first edition George Allen & Unwin, 1962; Boston: Houghton Mifflin, 1963, cited here from *Farmer Giles* (etc.), above

The Road Goes Ever On: A Song Cycle, Poems by J.R.R. Tolkien, Music by Donald Swann, Boston: Houghton Mifflin, 1967; London: George Allen & Unwin, 1968

Sir Gawain and the Green Knight, Pearl, Sir Orfeo, edited by Christopher Tolkien, London: George Allen & Unwin, 1975; Boston: Houghton Mifflin, 1975

'Guide to Names in *The Lord of the Rings*', revised for publication by Christopher Tolkien, in Jared Lobdell, ed., *A Tolkien Compass,* La Salle, Ill.: Open Court, 1975, 153–201

The Silmarillion, edited by Christopher Tolkien, London: George Allen & Unwin, 1977; Boston: Houghton Mifflin, 1977

Pictures by J.R.R. Tolkien, London: George Allen & Unwin, 1979; Boston: Houghton Mifflin, 1979, revised edition London: HarperCollins, 1992; Boston: Houghton Mifflin, 1992

Letters of J.R.R. Tolkien, edited by Humphrey Carpenter with the assistance of Christopher Tolkien, London: George Allen & Unwin, 1981; Boston, Houghton Mifflin, 1981 (cited in text as *Letters*)

Unfinished Tales of Númenor and Middle-earth, edited by Christopher Tolkien, London: George Allen & Unwin, 1980; Boston: Houghton Mifflin, 1980

The Monsters and the Critics and Other Essays, edited by Christopher Tolkien, London: George Allen & Unwin, 1983; Boston, Houghton Mifflin, 1984 (cited in text as *Essays*)

The History of Middle-earth, twelve volumes, all edited by Christopher Tolkien:

I, *The Book of Lost Tales, Part One,* London: George Allen & Unwin, 1983; Boston: Houghton Mifflin, 1984

II, *The Book of Lost Tales, Part Two*, London: George Allen & Unwin, 1984; Boston: Houghton Mifflin, 1984

III, *The Lays of Beleriand*, London: George Allen & Unwin, 1985; Boston: Houghton Mifflin, 1985

IV, *The Shaping of Middle-earth: The Quenta, the Ambarkanta, and the Annals*, London: George Allen & Unwin, 1986; Boston: Houghton Mifflin, 1986

V, *The Lost Road and Other Writings*, London: Unwin Hyman, 1987; Boston: Houghton Mifflin, 1987

VI, *The Return of the Shadow*, London: Unwin Hyman, 1988; Boston: Houghton Mifflin, 1988

VII, *The Treason of Isengard*, London: Unwin Hyman, 1989; Boston: Houghton Mifflin, 1989

VIII, *The War of the Ring*, London: Unwin Hyman, 1990; Boston: Houghton Mifflin, 1990

IX, *Sauron Defeated: The End of the Third Age, the Notion Club Papers and the Drowning of Anadune*, London: HarperCollins, 1992; Boston: Houghton Mifflin, 1992

X, *Morgoth's Ring: The Later Silmarillion, Part One*, London: HarperCollins, 1993; Boston: Houghton Mifflin, 1993

XI, *The War of the Jewels: The Later Silmarillion, Part Two*, London: HarperCollins, 1994; Boston: Houghton Mifflin, 1994

XII, *The Peoples of Middle-earth*, London: HarperCollins, 1996; Boston: Houghton Mifflin, 1996

The Old English Exodus, edited by Joan Turville-Petre, Oxford: Clarendon Press, 1981

Finn and Hengest: the Fragment and the Episode, edited by Alan Bliss, London: George Allen & Unwin, 1982; Boston: Houghton Mifflin, 1983

Tolkien's separately-published poems and academic works are listed in Carpenter, *Biography*, below, with further information in Hammond, *Bibliography*, below.

WORKS OF REFERENCE

Anderson, Douglas A., ed., *The Annotated Hobbit*, London: Unwin Hyman, 1988; Boston: Houghton Mifflin, 1988

Blackwelder, Richard A., *A Tolkien Thesaurus*, New York and London: Garland, 1990

Carpenter, Humphrey, *J.R.R. Tolkien: A Biography*, London: George Allen & Unwin, 1977; Boston: Houghton Mifflin, 1977 (cited in text as *Biography*)

Clute, John, and John Grant, eds., *The Encyclopedia of Fantasy*, New York: St Martin's, 1997

Hammond, Wayne, with the assistance of Douglas A. Anderson, *J.R.R. Tolkien: A Descriptive Bibliography*, Winchester: St Paul's Bibliographies; New Castle, Del.: Oak Knoll Books, 1993 (cited in text as *Bibliography*)

Hostetter, Carl F., http://www.elvish.org/resources.html

Johnson, Judith A., ed., *J.R.R. Tolkien: Six Decades of Criticism* (Bibliographies and Indexes in World Literature 6), Westport, Conn., and London: Greenwood, 1986

The Oxford English Dictionary, second edition prepared by J.A. Simpson and E.S.C. Weiner, 20 vols., Oxford and New York: Oxford University Press, 1989 (cited in text as *OED*)

FURTHER REFERENCES

Auden, W.H., 'At the End of the Quest, Victory', *New York Times Book Review*, January 22nd, 1956, p. 5

—— 'The Quest Hero', *Texas Quarterly* 4 (1961), 81–93, reprinted in Neil D. Isaacs and Rose A. Zimbardo, eds., *Tolkien and the Critics: Essays on Tolkien's 'The Lord of the Rings'*, Notre Dame and London: University of Notre Dame Press, 1968

Battarbee, Keith, ed., *Proceedings of the Tolkien Phenomenon* (Anglica Turkuensia 12), Turku, Finland: University of Turku Press, 1993

Carey, John, *The Intellectuals and the Masses: Pride and Prejudice among the literary intelligentsia, 1880–1939*, New York: St Martin's, 1993

Clark, George, and Dan Timmons, eds., *J.R.R. Tolkien and his Literary Resonances: Views of Middle-earth*, Westport, Conn. and London: Greenwood Press, 2000

Curry, Patrick, *Defending Middle-earth*, New York: St Martin's, 1997; London: HarperCollins 1998

—— 'Tolkien and his Critics: A Critique', in Honegger, ed., below, 81–148

Doughan, David, 'In Search of the Bounce: Tolkien seen through Smith', in *Leaves from the Tree*, below, 17–22

Drabble, Margaret, ed., *The Oxford Companion to English Literature*, Oxford and New York: Oxford University Press, 1998

Duggan, Alfred, 'Heroic Endeavour', *Times Literary Supplement*, August 27th, 1954, 541

—— 'The Epic of Westernesse', *TLS*, December 17th, 1954, 817

—— 'The Saga of Middle Earth', *TLS*, November 25th, 1955, 704

Flieger, Verlyn, *Splintered Light: Logos and Language in Tolkien's World*, Grand Rapids, Mich.: William B. Eerdmans, 1983

—— *A Question of Time: J.R.R. Tolkien's Road to Faerie*, Kent, Ohio: Kent State University Press, 1997

—— and Carl F. Hostetter, eds., *Tolkien's 'Legendarium': Essays on 'The History of Middle-earth'* (Contributions to the Study of Science Fiction and Fantasy 86), Westport, Conn., and London: Greenwood, 2000

Golding, William, 'Fable', in *The Hot Gates*, London: Faber & Faber, 1965, 85–101

Gordon, Ida L., ed., *Pearl*, Oxford: Clarendon Press, 1953

Graves, Robert, *Goodbye to All That*, first published London: Jonathan Cape, 1929, cited here from the Penguin Books edition, Harmondsworth: Penguin, 1960

Green, Martin, *Children of the Sun: A Narrative of 'Decadence' in England after 1918*, London: Constable, 1977

Greer, Germaine, 'The Book of the Century?', *W. The Waterstone's Magazine* 8 (Winter/Spring 1997), 2–9

Groden, Michael, and Martin Kreiswirth, eds., *Johns Hopkins Guide to Literary Theory and Criticism*, Baltimore, Md.: Johns Hopkins University Press, 1994

Hammond, Wayne, *J.R.R. Tolkien, Artist and Illustrator*, London: HarperCollins, 1995: Boston: Houghton Mifflin, 1995

—— '"A Continuing and Evolving creation": Distractions in the Later History of Middle-earth', in Flieger and Hostetter, above, 19–29

Hardy, James, ed., *The Denham Tracts*, London: Folklore Society, 2 vols. 1891–5

Honegger, Thomas, ed., *Root and Branch: Approaches towards Understanding Tolkien*, Zürich and Bern: Walking Tree Publishers, 1999

Hostetter, Carl F., and Arden R. Smith, 'A Mythology for England', in Reynolds and GoodKnight, below, 281–90

Leaves from the Tree: J.R.R. Tolkien's Shorter Fiction, no editor, London: Tolkien Society, 1991

Lewis, Alex, 'The Lost Heart of the Little Kingdom', in *Leaves from the Tree*, above, 33–44

Lewis, C.S., *The Screwtape Letters, with Screwtape Proposes a Toast*, New York: Macmillan, 1961

—— *Mere Christianity*, first published 1952, cited here from the Fontana paperback, London: Fontana, 1955

Manlove, Colin, *Modern Fantasy: Five Studies*, Cambridge: Cambridge University Press, 1975

Muir, Edwin, 'Strange Epic', *Observer*, August 22nd, 1954, 7

—— 'The Ring', *Observer*, November 21st, 1954, 9

—— 'A Boy's World', *Observer*, November 27th, 1955, 11

Noad, Charles, 'On the Construction of "The Silmarillion"', in Flieger and Hostetter, above, 31–68

Pearce, Joseph, *Tolkien: Man and Myth*, London: HarperCollins, 1998; San Francisco: Ignatius Press, 1998

Rateliff, John, '*The Lost Road, The Dark Tower*, and *The Notion Club Papers*: Tolkien and Lewis's Time-Travel Triad', in Flieger and Hostetter, above, 199–218

Reynolds, Patricia, and Glen H. GoodKnight, eds., *Proceedings of the J.R.R. Tolkien Centenary Conference*, Milton Keynes: Tolkien Society, and Altadena, Calif.: Mythopoeic Press, 1995

Roberts, Mark, *Essays in Criticism* 6 (1956), 450–9

Senior, William A., *Stephen R. Donaldson's 'Chronicles of Thomas Covenant': Variations on the Fantasy Tradition*, Kent, Ohio, and London: Kent State University Press, 1995

Shippey, T.A., *The Road to Middle-earth*, second edition, London: Grafton, 1992

—— 'Tolkien and 'The Homecoming of Beorhtnoth'', in *Leaves from the Tree*, above, 5–16

—— 'Tolkien as a Post-War Writer', in Battarbee, ed., above, 217–36, reprinted in Reynolds and Good Knight, eds., above, 84–93

—— 'Orcs, Wraiths, Wights: Tolkien's Images of Evil', in Clark and Timmons, eds., above, 181–96

—— ed., *The Oxford Book of Fantasy Stories*, Oxford and New York: Oxford University Press, 1994

Suvin, Darko, *Metamorphoses of Science Fiction: On the Poetics and History of a Literary Genre*, New Haven and London: Yale University Press, 1979

Thomas, Paul Edmund, 'Some of Tolkien's Narrators', in Flieger and Hostetter, above, 161–81

Tolkien, Christopher, ed. and trans., *The Saga of King Heidrek the Wise*, London: Nelson, 1960

Toynbee, Philip, 'The Writer's Catechism', *Observer*, April 23rd, 1961, 80

—— no title, *Observer*, August 6th, 1961

Unwin, Rayner, 'Publishing Tolkien', in Reynolds and Good Knight, eds., above, 26–9

White, T.H., *The Book of Merlyn*, Austin, Tx.: Texas University Press, 1977; London: William Collins, 1978

Wilson, Edmund, *Axel's Castle*, New York and London: C. Scribner's Sons, 1931

—— 'Oo, Those Awful Orcs!', *Nation* 182, April 14th, 1956, 312–13, reprinted in Wilson, *The Bit between my Teeth: A Literary Chronicle of 1950–1965*, New York: Farrer, Straus and Giroux, 1965, 326–32

Wynne, Patrick, and Hostetter, Carl F., 'Three Elvish Verse-Modes: *Ann-thennath*, *Minlamad thent/estent*, and *Linnod*, in Flieger and Hostetter, above, 113–39

Yates, Jessica, 'Macaulay and "The Battle of the Eastern Field"', *Mallorn*, 13 (1979), 3–5

—— 'The Source of "The Lay of Aotrou and Itroun"', in *Leaves from the Tree*, above, 63–71

—— 'Tolkien the Anti-totalitarian', in Reynolds and GoodKnight, above, 233–45

INDEX

Acton, Lord 115
Adam Bede (Eliot) 327
Adventures of Tom Bombadil, The
  (Tolkien) 25, 60, 265, 278, 280–1
'Adventures of Tom Bombadil, The'
  (poem, Tolkien) 60–1, 65, 269
Aeneid (Virgil) 121, 123
After the King (ed. Greenberg) 318
Alcuin 180, 182, 187, 259, 293
Alfred, King 131, 133–4
allegory viii, 160–8, 174, 187, 188,
  199, 204, 219, 289, 308
  autobiographical xxxiii, 265–77,
  296–304
Alvíssmál 20
America 284–5
anachronism xxviii–xxix, 6–7, 9, 11,
  23, 29, 36, 37, 39, 46–7, 60, 117,
  176, 235
Anatomy of Criticism (Frye) 221
Ancrene Wisse 270
Andersen, H.C. 12
Anderson, Douglas A. 24, 264, 265,
  278
angels 259–60
Anglo-Saxon Chronicle 92
Animal Farm (Orwell) vii, xxii, xxiii,
  115, 308
'Annals' (Tolkien) 228, 232, 236
Annotated Hobbit (Anderson) 24, 25
antiquarianism 86, 112, 159, 169
Antony and Cleopatra (Shakespeare)
  203
'Aotrou and Itroun' (Tolkien)
  293–4

applicability xxxi, 164–71, 173–4,
  188, 196, 205
Aragorn 72–3, 78
archaism xxxi, 18, 43, 69–70, 75,
  169, 224, 324
Arthurian legend 283, 284
Asbjørnsen & Moe 12, 22
Atlamál 22
Attila the Hun 176
Auden, W.H. 147, 192
Axel's Castle (Wilson) 307

Balrogs 85–7
Bard 39–40, 41, 45
Barfield, Owen 122
barrow-wights 61
barrows 61, 97, 175–6, 179
Battle of Maldon, The 92, 150–1,
  294
'Battle of the Eastern Field, The'
  (Tolkien) 236
Battle of the Goths and Huns, The
  34
'Bear and the Water-carl, The' 47–8
Bede's Death-Song 267
Beorn 31–2
Beowulf 28, 39, 62, 85, 88, 115, 146,
  169, 183, 283, 315
  Beowulf criticism 103, 162–3
  eucatastrophe in 215
  and fairy-tales 14, 47
  preliterate ancestry of 234–5
  as literary source 31, 32, 36, 42,
  49, 94–6, 99, 278, 286
  religion in 179–82

luck 27–8, 143–7, 155, 157, 173,
251, 325
Luther, Martin 210

Macaulay, Lord xxxii, 233, 234–6,
255, 262, 277, 289
Macbeth (Shakespeare) 192–4, 195,
254
MacDonald, George 12, 296
McKiernan, Dennis 318
magic 193–4, 196, 200
Maginot Line 166, 168, 174
Malory, Sir Thomas 158–9
Mandeville, Sir John 171
Manichaeanism 134–8, 141, 149,
157, 213, 295
Manlove, Colin 117, 118, 143
maps 58, 59, 65, 68, 102
Marie de France 293
Mark, the 91–2
Martin, George R.R. 326
Masson, David 134, 141
Maxims I 93
Meduseld 99–100
memorials 97, 155–6
Mercia 59, 91–2, 169
Mere Christianity (Lewis) 130, 131,
134, 258, 293
Middle English xii, 34, 89, 270
Midsummer Night's Dream, A
(Shakespeare) 196, 205, 211
Milton, John 17, 129, 130, 180, 186,
200, 202–4, 213, 285, 310, 314
mimesis, high and low 221–2, 223,
224, 256, 263
Minas Tirith 99–100, 213, 218
Mirkwood 33–4
Mr Bliss (Tolkien) 265
Modern Fantasy (Manlove) 117
modernism xvii, xxxi, 142, 171, 309,
312–18
*Monsters and the Critics and Other
Essays, The* (ed C. Tolkien) xvi,

149–50, 162, 179, 206, 211, 236,
240, 241, 291, 296, 299
Moore, G.E. 158
Moorman, Professor 63
morality 132–3, 147, 174
Morgoth's Ring (Tolkien) 228, 285
Muir, Edwin 147–8, 154, 155, 158,
159, 224
Murray, Robert 144
myth 85, 160, 221–3
Christian 179–80, 206, 212–14,
238–9
and fairy-tale 12, 13
as mediation of contradiction
179–82, 284
myth of Frodo 182–7
myth of stars and trees 202–6,
210
mythic timelessness xxxii, 188,
192, 196, 200, 210
Norse 150, 180
Tolkien's English mythology
231–3
Tolkien as mythologist xvi, xxxii,
201, 256, 313, 315
'Mythopoeia' (Tolkien) 293

'Nameless Land, The' (Tolkien) 197,
278, 279
names 15–17, 55, 57–60, 62, 84,
169–70, 182–4, 242, 303–4, 325
see also place-names
nationality 230, 231
see also Englishness
Neave, Jane 10, 57, 60
*New Glossary of the Dialect of the
Huddersfield District* (Haigh) 8
'New Lay of Sigurth, The' (Tolkien)
278
New York Times xiii
Nineteen Eighty-Four (Orwell) vii,
xxii, xxiii, xxxi, 119, 168
Noad, Charles 228